Study Guide

for Bettelheim, Brown, & March's

Introduction to General, Organic, and Biochemistry

Seventh Edition

William M. Scovell
Bowling Green State University

THOMSON
BROOKS/COLE

Australia • Canada • Mexico • Singapore • Spain • United Kingdom • United States

Printed in the United States of America
2 3 4 5 6 7 07 06 05 04 03

Printer: Patterson Printing Company
Image: © FoodPix/Karl Petzke

0-534-40178-3

For more information about our products, contact us at:
Thomson Learning Academic Resource Center
1-800-423-0563

For permission to use material from this text, contact us by:
Phone: 1-800-730-2214
Fax: 1-800-731-2215
Web: http://www.thomsonrights.com

Asia
Thomson Learning
5 Shenton Way #01-01
UIC Building
Singapore 068808

Australia/New Zealand
Thomson Learning
102 Dodds Street
Southbank, Victoria 3006
Australia

Canada
Nelson
1120 Birchmount Road
Toronto, Ontario M1K 5G4
Canada

Europe/Middle East/South Africa
Thomson Learning
High Holborn House
50/51 Bedford Row
London WC1R 4LR
United Kingdom

Latin America
Thomson Learning
Seneca, 53
Colonia Polanco
11560 Mexico D.F.
Mexico

Spain/Portugal
Paraninfo
Calle/Magallanes, 25
28015 Madrid, Spain

TO THE STUDENT

I hear, and I forget
I see, and I remember
I do, and I understand

"Ancient Chinese Proverb"

The aim of this Study Guide is simply to aid the student in increasing understanding of the material presented in Introduction to General, Organic and Biochemistry by Bettelheim, Brown & March, and to make studying more efficient. The general philosophy or approach put forth here should not only be useful in this course, but in many courses in which the learning of specific new concepts and the development of problem solving abilities is important. The initial focus of attention should be to determine the immediate objectives or aims of each chapter. Therefore study objectives will begin every chapter in order to point out aspects worthy of special emphasis in your studies.

Once the major objectives have been read over carefully, one must ask, "How best can I study to develop a firm understanding and a command of the new vocabulary, concepts and problem solving techniques associated with the material?" The approach to this end will be to define four study practices.

1. Each chapter in the text includes a summary of the key concepts in that chapter. It is essential that the meaning of these terms, in **bold type**, be mastered. These and additional terms of importance from the text will be listed in the Study Guide, either individually, or in cases in which two or more terms are related, will be listed together. This latter grouping will serve to emphasize that the similarities and differences between these terms should be clearly understood. The terms are basic to the understanding of the new material and will provide the very foundation of your understanding. Review them together with the objectives before reading the chapter. After reading the text, reread the objectives and define each term. To deepen your understanding of the terms, include a specific example in each case where this is appropriate. Refer to the Glossary in the text to provide clear definitions to all the terms. **Continually review** the definitions before each study session or until the responses are clear and spontaneous.

2. In each chapter, a few concepts provide the basis for grasping many of the sometimes apparently diverse, but interrelated subjects presented. In some Study Guide chapters, a selected number of concepts will be elaborated on in order to "pull together" some of the ideas and to deepen understanding of the text material.

3. An integral part of chemistry involves the solving of both mathematical and conceptual problems. The general thought processes and interrelationships helpful in reaching these solutions will be briefly outlined. These problems will, in some cases, include the detailed solutions of some of the problems in the text.

4. At the point at which the student feels confident that the material has been mastered, a **SELF-TEST** will provide a means for evaluating the progress toward reaching the major objectives in the chapter. Answers to all the questions and problems will be included, with explanations accompanying some of the answers.

SOME OBSERVATIONS AND A SUGGESTED STUDY APPROACH

1. Typically students who obtain above average grades will read the chapter a minimum of twice, if not three times. I point it out only because I find many freshman (and some upper classmen also) are surprised to learn that this effort is actually the **norm** and to be sure, not unusual.

2. Reading the chapter prior to the material being discussed in class usually affords the student the unique advantage of being able to clarify questions on the material at the time the professor is introducing them to the class. Unfortunately, for many students, this "pre-reading" often becomes very difficult, if not impossible. However, try it. You may find it so valuable that it may become a part of your personal study strategies.

3. Take good notes in class and as soon after class as possible, **recopy the notes into another notebook in a more organized fashion**. Do this at the same time that you refer to your textbook and annotate your class notes with extra definitions or comments from the text. I have found that this has been especially helpful to many students.

4. Knowledge of a subject begins with **memorizing** the meaning of new terms, concepts and equations. However, **understanding** the material goes far beyond the memorization of these elements in a "vacuum." It requires that the interrelationships between all these elements be clearly resolved. This requires a continual review of the material and a major input of time and effort into solving problems and answering focused questions. Therefore, it is important that you commit yourself to working as many problems as possible. There is no substitute for this.

I would like to outline a study procedure that many students have found helpful. This, of course, will not be for everyone. Each student must develop routines which serve him or her best. Try this approach if you wish. If it helps, use it.

1. **Review** the chapter objectives and important new terms in the Study Guide. This will alert you to the key ideas and concepts and help to underline their importance before your reading.

2. **Read** the (text) chapter with these points in mind.

3. **Record** good notes in class and as soon after class as possible, rewrite the notes. Clarify all questionable points in the lecture notes **AT THIS TIME**, with the aid of your text. For those concepts which are not clear, **ask the professor immediately** after the next lecture or bring up this point at his or her next office hour.

4. **Write out** the definitions of all important new terms.

5. **Review** any worked problems in the chapter, with the aim of defining the general thought processes involved in solving the problem.

6. Now, the assigned questions and problems at the end of each chapter in the text should be worked and the material in the Study Guide should be studied. The guide will be most useful in cases in which some additional explanation or worked examples are helpful. In some cases, simply rephrasing of text material may bring out the meaning more clearly.

7. **Reread** the chapter, the rewritten class notes, chapter objectives, important terms and equations and the focused review section.

8. Return to doing questions and problems that were initially unclear or difficult and then do additional ones.

9. **Do the SELF-TEST** in the study guide.

TABLE OF CONTENTS

Chapter	Title	Page
Part 1:	**General Chemistry**	
1	Matter, Energy and Measurement	1
2	Atoms	21
3	Chemical Bonds	26
4	Chemical Reactions	34
5	Gases, Liquids, and Solids	46
6	Solutions and Colloids	56
7	Reaction Rates and Chemical Equilibrium	63
8	Acids and Bases	69
9	Nuclear Chemistry	76
Part 2:	**Organic Chemistry**	
10	Organic Chemistry	83
11	Alkanes	89
12	Alkenes	96
13	Benzene and Its Derivatives	107
14	Alcohols, Ethers, and Thiols	113
15	Chirality: The Handedness of Molecules	122
16	Amines	129
17	Aldehydes and Ketones	136
18	Carboxylic Acids, Anhydrides, Esters, and Amides	146

Part 3: Biochemistry

19	Carbohydrates	157
20	Lipids	164
21	Proteins	170
22	Enzymes	178
23	Chemical Communications: Neurotransmitters and Hormones	185
24	Nucleotides, Nucleic Acids, and Heredity	191
25	Gene Expression and Protein Synthesis	199
26	Bioenergetics: How the Body Converts Food to Energy	206
27	Specific Catabolic Pathways: Carbohydrate, Lipid, and Protein Metabolism	216
28	Biosynthetic Pathways	221
29	Nutrition	226
30	Immunochemistry	231
31	Body Fluids	235

"It is only the first step
which takes the effort"
- Madame Marie Vichy-Deffand

1

Matter, Energy and Measurements

CHAPTER OBJECTIVES

After studying Chapter 1 and the material in the Appendix on significant figures, and working the assigned exercises in the text and the study guide, you should be able to do the following.

1. Distinguish between a fact, a hypothesis and a theory. Indicate how they are associated with the scientific method.

2. Convert large and small numbers to an equivalent form in exponential notation. Carry out the reverse procedure.

3. Perform addition, subtraction, multiplication and division calculations using exponential notation. Evaluate the answers to insure that there are the proper number of significant figures.

4. List the word prefixes used in the metric system and the corresponding values.

5. Define temperature and indicate temperature scales in common use. Write the equations which relate temperature in (i) ^{o}C to ^{o}F; (ii) ^{o}F to ^{o}C (iii) ^{o}C to K; perform temperature conversions from one scale to another.

6. Using the factor-label method, perform conversions between the metric and English units, English to metric units, in addition to conversions within the metric system.

7. List the three states of matter and the characteristics which are unique to each state.

8. Distinguish between density and specific gravity.

9. Define energy and list six different forms of energy pointed out in this chapter.

10. State the Law of Conservation of Energy.

11. Define heat and specific heat and indicate the units of each.

12. Using the rules for determining the number of significant figures, evaluate the number of significant figures in numbers listed throughout the chapters.

13. Define the important terms and comparisons in this chapter and give specific examples where appropriate.

IMPORTANT TERMS AND COMPARISONS

Matter and Energy
Fact, Hypothesis and Theory
Coefficient and Exponent
Kilo- and Milli-
Mega- and Micro
Centi-
Mass and Weight
Significant Figures
Scientific Method
Chemical and Physical Properties
Chemistry
Centigrade, Fahrenheit and Kelvin Scale
States of Matter
Density and Specific Gravity
English vs. Metric System of Units
Potential and Kinetic Energy
Law of Conservation of Energy
Temperature and Heat
Calorie and Joule
Hypothermia and Hyperthermia
Exponential Notation
Measured and Defined Numbers

FOCUSED REVIEW OF CONCEPTS

A. Interrelationships

(i) $°F = 9/5(°C) + 32$

(ii) $°C = 5/9(°F-32)$

(iii) $K = °C + 273$

(iv) Density = mass/volume; $D = M/V$
D is a **physical quantity** and has **units**

(v) Specific Gravity (S.G.) =

$$\frac{\text{Density of substance of interest}}{\text{Density of water}}$$

$S.G. = D_A/D_{H_2O}$

S.G. is a unitless number

(vi) Heat (cal) = Specific heat (cal/g-deg) x mass (g) x (T_2-T_1)

It should be noted that both equations (i) and (ii) define the relationship between the temperature in degrees Centigrade and in degrees Fahrenheit. Therefore, you should be able to rearrange equation (i) to arrive at equation (ii).

Write eqn (i)

$°F = 9/5\,(°C) + 32$

Subtract 32 from each side

$$°F-32 = 9/5(°C)+32-32$$

$$(°F-32) = 9/5(°C)$$

Multiple each side by 5/9

$$5/9(°F-32) = (5/9)(9/5)(°C)$$

This is eqn (ii)

$$5/9(°F-32) = °C$$

Equation (iii) should be put to memory since many calculations to be performed in subsequent chapters (e.g., Chapter 5) will require that the temperature be expressed in degrees Kelvin (K).

Density is an intensive property of a substance and has units of mass/vol, which although usually expressed in g/mL, can be expressed in others units (i.e., lb/in^3, etc.). Recall also that **intensive** properties are those which are **int**rinsic characteristics of the substance and therefore do not depend on the quantity of the substance.

Specific gravity is a ratio of the density of any substance to the density of water at 20°C. Therefore, specific gravity is simply a number, with no units.

B. Significant Figures

A scientist must evaluate the number of significant figures in every measurement and in every calculation using measured numbers. The number of significant figures is defined as the number of digits that are known with certainty.

The rules are the following.

1. All digits written down are significant, except in the case of zeros.
   a. Zeros that come before the first non zero digit are not significant.
   b. Zeros that come after the last non zero digit are significant if the number is a decimal. However, if the number is a whole number (no decimal point), they may or may not be significant.
2. In multiplying or dividing numbers, the **final answer** should have the same number of significant numbers as there are in the number with the fewest significant numbers.
3. In adding or subtracting a group of numbers and arriving at the number of significant figures in the answer, the number of significant numbers in the individual numbers does not matter. The answer must contain the same number of decimal places as the number in the listing with the fewest decimal places.
4. In rounding off numbers, if the first digit dropped is a 5, 6, 7, 8 or 9, raise the last digit kept to the next higher number. In all other cases, do not raise the last digit.
5. Counted or defined numbers, such as the number of sides on a square, are treated as if they have an infinite number of zeros following the decimal point.

PROBLEM SOLVING METHODS

A. Exponential Notation

It is important to be able readily to express large and small numbers in exponential or scientific notation. This requires that the number be written as a **coefficient,** with a value between 1 and 10, **multiplied** by 10 raised to some power. For example,

$\underline{100} = 1 \times 10^2 = 1 \times (1\underline{00})$ — The solid line underlines the number of places the decimal point has been moved in each case.

$2\underline{0,000} = 2 \times 10^4 = 2 \times (1\underline{0,000})$

$5\underline{,200,000} = 5.2 \times 10^6 = 5.2 \times (1\underline{,000,000})$

Note in these examples that when the LARGE number (greater than 1) is changed to exponential notation form, the decimal point is moved to the LEFT by 2, 4 or 6 places, respectively. The positive exponent simply expresses the number of places that the decimal has been moved to the left.

$0.\underline{001} = 1 \times 10^{-3} = 1 \times (0.\underline{001})$

$0.\underline{00004}3 = 4.3 \times 10^{-5} = 4.3 \times (0.\underline{00001})$

$0.\underline{02}00100 = 2.001 \times 10^{-2} = 2.001 \times (0.\underline{01})$

Similarly, in changing these SMALL numbers (less than one) to exponential notation, the decimal point is moved to the RIGHT by 3, 5 or 2, respectively.

a. MOVING THE DECIMAL POINT in Numbers Already in Exponential Notation.

One of the most common difficulties that students encounter occurs when doing multiple step calculations. Midway through the calculation, numbers such as the following may appear.

$16,743 \times 10^{14}$ $\quad 0.00713 \times 10^{6}$ $\quad 0.0000431 \times 10^{-6}$

To change these numbers to proper form (i.e., the coefficient being between 1 and 10), the decimal point must be moved. Which way is always the question!

The rule to remember is:

IF THE **COEFFICIENT IS DECREASED** in magnitude, the **EXPONENT MUST BE INCREASED ACCORDINGLY.**

IF THE COEFFICIENT IS INCREASED, the **EXPONENT IS DECREASED.**

For example:

$1\underline{6,743} \times 10^{14} \rightarrow 1.6743 \times 10^{18}$

coefficient decreased, decimal moved **4** places
exponent increased by 4

$0.\underline{007}13 \times 10^{6} \rightarrow 7.13 \times 10^{3}$

coefficient increased, decimal moved **3** places
exponent decreased by 3

$0.\underline{0000}431 \times 10^{-6} \rightarrow 4.31 \times 10^{-11}$

coefficient increased, decimal moved **5** places
exponent decreased by 5

Exercise I: Carry out these manipulations to develop skill in these conversions.

Number	Exponential Notation (with coefficient between 1 and 10)
1. 0.00102×10^{1}	____________
2. 632×10^{-14}	____________
3. 0.00700×10^{43}	____________
4. $13{,}642{,}000 \times 10^{-1}$	____________
5. 11×10^{23}	____________
6. $0.0000006 \times 10^{-14}$	____________
7. 137×10^{-10}	____________
8. 0.006×10^{0}	____________
9. $7{,}423{,}000{,}000 \times 10^{17}$	____________
10. 0.4361×10^{31}	____________

In the frequently used manipulations with numbers in scientific notation outlined below, a TWO-STEP procedure can be used in each case.

b. ADDITION OR SUBTRACTION in Exponential Notation

Procedure: (i) change the numbers so all have the same exponent, irrespect of the size of the coefficient
(ii) Add (or subtract) the coefficients; the exponent remains the same

Ex. a. 230×10^{-8}, $\underline{6.10 \times 10^{-5}} \rightarrow$ (i) 23×10^{-7}, $\underline{610 \times 10^{-7}}$

(ii) ↓ ↓

$633 \times 10^{-7} = 6.33 \times 10^{-5}$

b. 340×10^{6}, $- \underline{220 \times 10^{5}}$ (ii) → 340×10^{6}, $- \underline{22 \times 10^{6}}$

(ii) ↓ ↓

$318 \times 10^6 = 3.2 \times 10^8$

c. **MULTIPLICATION** in Exponential Notation

Procedure:
(i) Multiple the coefficients directly
(ii) Algebraically **add** the exponents
(iii) Combine coefficients and exponent
(iv) If necessary, change the coefficient to a value between 1 and 10 and make the corresponding change in the value of the exponent.

Ex. a. $(6.0 \times 10^{-14})(3.0 \times 10^{6})$
(i) → 18
(ii) → $\times 10^{-8}$
(iii) 18×10^{-8}
(iv) $\underline{1.8 \times 10^{-7}}$

b. $(1 \times 10^{-14})(7 \times 10^{-2})$
(i) → 7
(ii) → $\times 10^{-16}$
(iii & iv) $\underline{7 \times 10^{-16}}$

d. **DIVISION** in Exponential Numbers

Procedure:
(i) Divide the coefficients directly
(ii) Algebraically subtract the exponents
(iii) Combine
(iv) If necessary, change coefficient to a value between 1 and 10 and make the corresponding change in the value of the exponent.

Ex. a. $\dfrac{3 \times 10^{-7}}{1 \times 10^{7}}$
(i) → 3
(ii) → $\times 10^{-7-7}$
(iii & iv) $\underline{3 \times 10^{-14}}$

b. $\dfrac{3 \times 10^{7}}{1 \times 10^{-7}}$
(i) → 3
(ii) → $\underline{\times 10^{7-(-7)}}$
(iii & iv) $\underline{3 \times 10^{14}}$

c. $\dfrac{2.14 \times 10^{-3}}{8.60 \times 10^{+4}}$
(i) → 0.249
(ii) → $\underline{\times 10^{-3-4}}$
(iii & iv) $0.249 \times 10^{-7} = \underline{2.49 \times 10^{-8}}$

d. $\dfrac{5.91 \times 10^{-4}}{1.61 \times 10^{-12}}$
(i) → 3.67
(ii) → $\underline{\times 10^{-4+12}}$
(iii & iv) $\underline{3.67 \times 10^{8}}$

B. Unit Conversions

(i) All conversions within the metric system differ by some power of 10 as reflected in the prefixes. For example, centi = 10^{-2}; milli = 10^{-3}; micro = 10^{-6}; nano = 10^{-9}. These conversions are accomplished simply by moving the decimal point. Memorize these prefixes and those in Table 1.2 (text).

The strategy in these problems is to multiply the starting physical quantity (a number with a unit) by a conversion factor which will both change the number and also its units.

AT EACH STEP IN THE CONVERSION, ONE UNIT MUST CANCEL OUT. AT NO POINT IN THE PROCESS ARE THERE MORE UNITS THAN YOU STARTED WITH. This conversion process is continued until the desired unit (or units) is (are) obtained. In each case, clearly understand what the initial units are and also what the units are you want to convert to.

Ex. 1 Change 2 ft to inches. Start with **feet;** Convert to **inches**

$$\left(\frac{2\ \cancel{ft}}{1}\right)\left(\frac{12\ in}{1\ \cancel{ft}}\right) \rightarrow \frac{24\ in}{1} = 24\ \text{inches}$$

starting physical quantity — conversion factor

Note that in this one-step conversion, 2 units cancel (ft) and only one unit (in., the desired unit) remains. If the correct conversion factor had been used, but incorrectly inverted, **no units would have canceled**. For example:

$$\left(\frac{2\ ft}{1}\right)\left(\frac{1\ ft}{12\ in}\right) \rightarrow \frac{2\ ft^2}{12\ in}$$

Units did not cancel. This incorrect set-up produced more units (ft^2/in) than in the starting quantity (ft). This is your signal that the conversion factor is inverted.

Ex. 2 Convert 3.2 milliliters to nanoliters.

Start with **milliliters**; want **nanoliters.** Depending on the conversion factors you know, this problem and many others, can be done in one or perhaps a few steps.

Route 1: $\left(\frac{3.2\ \cancel{mL}}{1}\right)\left(\frac{10^6\ nL}{1\ \cancel{mL}}\right) = 3.2 \times 10^6\ nL$

Route 2: $\left(\frac{3.2\ \cancel{mL}}{1}\right)\left(\frac{10^3\ \cancel{\mu L}}{1\ \cancel{mL}}\right)\left(\frac{10^3\ nL}{\cancel{\mu L}}\right) = 3.2 \times 10^6\ nL$

Although 1 nL = 10^{-9} L = 10^{-6} mL = 10^{-3} μL, you should note that the conversion factors are:

$$\frac{10^9\ nL}{L},\ \frac{10^6\ nL}{mL},\ \frac{10^3\ nL}{\mu L}.$$

In each case, the exponents are **positive** to show the number of nanoliters contained in the larger volume unit.

Exercise II: Convert the quantities shown below into a metric equivalent. Since all the quantities are in the metric system, the conversion requires only moving the decimal to the left or the right.

1.	1.32 L	______ mL	______ μL	______ nL
2.	6.32 mg	______ g	______ kg	______ Mg
3.	0.039 cm	______ m	______ km	______ μm
4.	0.36 μs	______ ms	______ s	______ ns

(ii) General conversions that are not within the metric system require the same procedure. The conversion factors used, however, do not simply move the decimal point (i.e., the exponent), but change both the coefficient and the exponent in the starting number.

Ex. 1 How many nickels are in $363.00?
Start with **dollars**; convert to **nickels**.

$$\frac{363\ \text{dollars}}{1}\ \frac{20\ \text{nickels}}{1\ \text{dollar}} = (363)(20)\ \text{nickels} = 7{,}260\ \text{nickels}$$

Ex. 2 The density of water is 1 g/cc. What is this equivalent to in terms of lb/in^3?

Solution:

This is a multistep conversion in which 2 units are to be converted.

g → lb.

cc. → in^3.

These problems are done by **completely converting one unit**, and then the other.

Recall 1 cc = 1 cm^3

$$\left(\frac{1.0\ \text{g}}{\text{cm}^3}\right)\left(\frac{1.0\ \text{lb}}{454\ \text{g}}\right)\left(\frac{2.54\ \text{cm}}{1\ \text{in}}\right)\left(\frac{2.54\ \text{cm}}{1\ \text{in}}\right)\left(\frac{2.54\ \text{cm}}{1\ \text{in}}\right) = \frac{(2.54)^3\ \text{lb}}{454\ \text{in}^3}$$

conversion from g to lb

conversion of $cm^3 \rightarrow in^3$

$$= 3.6 \times 10^{-2}\left(\frac{\text{lb}}{\text{in}^3}\right)$$

Ex. 3 Which of the following is the largest mass, and which is the smallest?

a) 3.5×10^{-8} kg

b) 27 g

c) 4×10^3 mg

d) 3.4×10^6 μg

Solution:

The main difficulty in determining which quantity is the largest and which is the smallest occurs because each mass is expressed in a different unit. The first task then is to convert all the masses to the same unit. It is not important **which** units are finally compared (kg. with kg., g with g or other units). The question here will be answered by comparing the masses in units of μgs.

a. $\left(\frac{3.5 \times 10^{-8}\ \text{kg}}{1}\right)\left(\frac{10^{3}\ \text{g}}{1\ \text{kg}}\right)\left(\frac{10^{6}\ \mu\text{g}}{1\ \text{g}}\right) = 3.5 \times 10^{-8+3+6}\ \mu\text{g}$

$= 3.5 \times 10^{1}\ \mu\text{g}$

b. $\left(\frac{27\ \text{g}}{1}\right)\left(\frac{10^{6}\ \mu\text{g}}{1\ \text{g}}\right) = 2.7 \times 10^{7}\ \mu\text{g}$

c. $\left(\frac{4 \times 10^{3}\ \text{mg}}{1}\right)\left(\frac{10^{3}\ \mu\text{g}}{1\ \text{mg}}\right) = 4 \times 10^{6}\ \mu\text{g}$

d. $3.4 \times 10^{6}\ \mu\text{g} = 3.4 \times 10^{6}\ \mu\text{g}$

The comparison is now between all masses expressed in μg.

a. 3.5×10^{1} μg

b. 2.7×10^{7} μg or 27×10^{6} μg

c. 4×10^{6} μg

d. 3.4×10^{6} μg

$3.5 \times 10^{+1}\ \mu\text{g} < 3.4 \times 10^{6}\ \mu\text{g} < 4 \times 10^{6}\ \mu\text{g} < 2.7 \times 10^{7}\ \mu\text{g}$

increasing mass →

Therefore,

a. has the smallest mass (35 μg)

b. has the largest mass (2.7×10^{7} μg or 27,000,000 μg)

Work this problem again for practice and convert all the masses to a different common unit.

Exercise III: Perform the following conversions from the English to the metric system of units or vice versa. These conversions require the use of conversion factors and the factor label method. Refer to Table 1.3 (text) for the less widely used conversion factors.

1.	1.0 ft	________in	________cm	________mile
2.	300 g	________oz(avdp)	________lb	________µg
3.	6.3 fluid oz.	________µL	________L	________qt

Ex.4 If the density of air is 1.25×10^{-3} g/cc, what is the mass of the air in a room that is 5.0 meters long, 4.0 meters wide and 2.2 meters high?

Solution:

One approach to this problem is the following:

(i) The density is known in g/cc.

(ii) Determine the volume of the room in meters cubed (m^3) and then convert this volume to cc.

(iii) Knowing the density now in g/cc and the volume in cc, calculate the mass of the air in the room in grams.

(i) $D = 1.25 \times 10^{-3}$ g/cc

(ii) Determine the volume

Volume = length x width x depth = (5.0 m) (4.0 m) (2.2 m) = 44 m^3

Conversion of 44 m^3 to cc

$$\left(\frac{44\ \cancel{m^3}}{1}\right)\left(\frac{10^2\ cm}{1\ \cancel{m}}\right)\left(\frac{10^2\ cm}{1\ \cancel{m}}\right)\left(\frac{10^2\ cm}{1\ \cancel{m}}\right) = 44 \times 10^6\ cm^3 \text{ or } 4.4 \times 10^7\ cc$$

(iii) Determine the mass of air.

$$D = \frac{M}{V} \qquad M = (D)(V)$$

$$= (1.25 \times 10^{-3}\ g/cc)(4.4 \times 10^7\ cc)$$

$$M = 5.5 \times 10^4\ g$$

THE CONCEPT OF SPECIFIC HEAT

The specific heat of a substance is an intensive property of a substance - that is, it does not depend on the quantity of the substance. It is a measure of how much energy (heat) is required to increase the temperature of one gram of a substance by 1°C.

Ex. 1 Consider the specific heat of 3 substances.

specific heat (cal/g-deg)	
water	1.00
iron	0.11
lead	0.038

a. Which substance requires the most heat to increase 1 g of the substance 1°C?

b. If equivalent amounts of heat (for ex., 100 cal) are added to 1 g of each substance, which substance will experience the largest increase in temperature? Which will experience the least increase?

c. If 60 cals. are added to 6 g of iron at 20°C, what will the final temperature be?

Solution:

a. Using the definition of specific heat directly, the values indicate that the specific heat for water requires 1.00 cal to increase 1 g of water by 1°C. The specific heat value for iron is about one tenth the value for water, while the value for lead is smaller yet. Therefore, by definition and simple comparison, water requires the most heat.

b. An equivalent amount of heat added to 1 g of each substance would increase the temperature of lead the most, while the increase in water would be the least. This is exactly the opposite order as observed in part a.

c. **Heat** absorbed by a substance will produce an increase in **temperature**. The magnitude of the increase will depend on the **specific heat** (S.H.) of the substance and the **mass**. As shown in the section 1.9 in the text, the equation relating these quantities is:

$$\text{heat} = (\text{S.H.}) \times (m) \times (T_2 - T_1)$$

This equation has 5 quantities (heat, S.H., m, T_2, T_1). In this problem and in others, you must first determine (i) which quantities are given and (ii) which quantity is to be determined (i.e., the unknown). Since only one quantity in this problem is to be determined, 4 quantities must be known.

Known quantities:

heat = 60 cal

m = 6 g

S.H. (iron) = 0.11 cal/g-deg

T_1 = 20°C

T_2 = to be determined

Inserting these quantities into the equation

$$60 \text{ cal} = (0.11 \text{ cal/g-deg})\ (6g)\ (T_2\text{-}20°)$$

Rearrange this equation by dividing both sides by (0.11 cal/g-deg) (6 g) and then canceling on the right side to yield

$$\frac{60 \text{ cal}}{(0.11 \text{ cal/g-deg})\ (6 \text{ g})} = (T_2\text{-}20°C)$$

$$90° = T_2\text{-}20$$

Finally, add 20 to both sides to arrive at the final temperature, T_2.

$$110°C = T_2$$

Ex. 2 If 425 cal of heat are added to a sample of 52 g of copper at 25°C, what will be the final temperature?

$$\text{heat} = (\text{S.H.}) \times (m) \times (T_2\text{-}T_1)$$

Given:

heat	=	425 cal
m	=	52 g
S.H. (copper)	=	0.092 (Table 1.3, text)
T_1	=	25°C
T_2	=	to be determined

$$425 \text{ cal} = 0.092 \text{ (cal/g-deg)} \times (52 \text{ g}) \times (T_2\text{-}25°C)$$

$$\frac{425 \text{ cal.}}{(0.092 \text{ cal/g-deg})\ (52 \text{ g})} = T_2\text{-}25°C$$

$$89° = T_2\text{-}25°C$$

$$114° = T_2$$

SELF-TEST QUESTIONS

MULTIPLE CHOICE. In the following questions, select the correct answer from the choices listed. In some cases two or more answers will be correct.

1. Indicate which of the following are extensive properties of a substance - that is, a property that depends on the quantity of the substance.

a. mass
b. density
c. length
d. specific heat
e. kinetic energy
f. melting point

2. Indicate whether each statement is a fact, hypothesis or a theory.

a. Smoking cigarettes is detrimental to your health.
b. The ocean water at Miami, Florida contains more salt than the water in Galvaston, Texas.
c. A micrometer (um) is larger than a nanometer (nm).
d. Judy received a higher grade on the exam because she studied longer.

3. Which of the following are forms of matter?

a. thought
b. nucleus of an atom
c. a vacuum
d. a line drawn on a paper
e. a shadow

4. The number 2.2×10^{-4} is equivalent to:

a. 0.0022
b. 22×10^{-5}
c. 22×10^{-3}
d. 0.022×10^{-2}

5. The number 1×10^{-4} is equivalent to:

a. $\frac{1}{1 \times 10^{4}}$
b. 10×10^{-3}
c. 0.0001
d. 100×10^{-6}

6. Which of the following numbers can be added directly in exponential notation?

a. 3.26×10^{-4}
b. 600×10^{-2}
c. 32.6×10^{-3}
d. 0.40×10^{-4}

7. Multiplication of $(3 \times 10^{-7})\ (1 \times 10^{6})\ (2 \times 10^{0})$ equals:

a. 3×10^{-1}
b. 6×10^{-1}
c. 6
d. 60×10^{-3}

8. The answer to the problem $\frac{93 \times 10^{4}}{3.0 \times 10^{-4}}$ is:

a. $31 \times 10^{+8}$
b. 31
c. 13
d. 31×10^{0}

9. The answer to the problem $\frac{(3 \times 10^{-7})(5 \times 10^{2})}{3 \times 10^{1} + 20}$ is:

a. 5×10^{-4} c. 3×10^{-4}
b. 3×10^{-6} d. 0.005

10. Which of the following are equivalent to 50 mL?

a. 0.5L c. 0.05 L
b. 50 cc d. 5×10^{4} μL

11. Which of the following are equivalent to .5 ms?

a. 500 μs c. 8.3×10^{-6} min
b. 5×10^{-4}s d. 5000 μs

12. A degree on the Centigrade scale is:

a. larger than the degree on the Fahrenheit scale
b. is the same size as the degree on the Kelvin scale
c. 9/5 larger than the Fahrenheit scale
d. none of the above

13. The temperature of -40°C corresponds to:

a. 313 K c. -40°F
b. 233 K d. -73°F

14. The temperature of 0 °K corresponds to:

a. 273°C c. -459.4°F
b. absolute zero d. 32°F

15. A density of 1.5 g/cc is equivalent to:

a. 1500 g/L c. 3.3×10^{-3} lb/mL
b. 13.4 lb/ft^{3} d. 6

16. How much energy is required to warm 15.0 g of ethyl alcohol from 4°C to 25°C? The specific heat of ethyl alcohol is 0.590 cal/g-deg.

a. 158 cal c. 0.185 kcal
b. 185 cal d. 37 cal

17. Which of the following ratios may be used directly as a conversion factor in converting microliters (μL) to liters (L)?

a. $\frac{1 \text{ gal}}{3.785 \text{ L}}$ b. $\frac{1\text{L}}{10^{+6} \text{ uL}}$ c. $\frac{10^{6} \mu\text{L}}{1 \text{ L}}$ d. $\frac{10^{-6} \text{ uL}}{1\text{L}}$

18. Indicate the number of significant figures in the following measured numbers.

a. 0.02 b. 32.00032 c. 0.00011 d. 120,000

COMPLETION. Write the word, phrase or number in the blank that will complete the statement or answer the question.

1. Matter is defined as ______________________________.

2. Scientists proposed new ____________ to be tested and continually examine the validity of ____________ in more depth.

3. In the number, 4×10^{-4}, the number 4 is called the ____________, while -4 is the ____________.

4. Another way of writing the following numbers is:

a. $\frac{1}{10^{-4}}$ = ____________ c. 43×10^{6} = ______________

b. 0.001 = ____________ d. 9,430,000,000 = ______________

5. The number $87{,}000 \times 10^{-16}$ is the same as 8.7×10^{-n} in which n equals ______________.

6. SI is the abbreviation for ______________. This system is based on the ______________.

7. In the SI, the basic unit of length is the ______________, which is abbreviated as __________; the basic unit of volume is ____________, which is abbreviated as ____________.

8. A kilometer is ________ times longer than a centimeter.

9. Scientists use a balance to experimentally determine the __________ of a substance.

10. The basic unit for ____________ is the same in the English and metric systems. This abbreviation is ____________.

11. The size of the degree on the Kelvin scale is _________ to that on the Centigrade scale.

12. A temperature of 300°C corresponds to ________ K.

13. A temperature of 37°C is equal to _______ °F.

14. Absolute zero is defined as _________ K. At this temperature, the motion of molecules ____________.

15. There are _________ L in 6.7 gallons of wine.

16. The three states of matter are __________, __________ and ____________.

17. To change a liquid to a gas, _____________ must be added to the liquid.

18. To change a liquid to a solid, energy must be __________ from the liquid.

19. For the three states of matter, the order of compressibility is ________ > ________ > ________.

20. Oil spilled by large tankers floats on the ocean because oil is ________ than ocean water.

21. The density and the specific gravity of a substance differ in that ____________.

22. Density and specific gravity are ____________ properties of a substance.

23. The density of a substance ____________ as the temperature is increased.

24. A hydrometer is used to measure the ____________ of a liquid.

25. Energy is defined as ________________.

26. Kinetic energy is the energy of ______________, while potential energy is ____________ energy.

27. The law of conservation of energy states that ______________________________.

28. A high value for the specific heat of a substance indicates that __________________.

29. Substance 1 has a specific heat of 3.0 cal/g-deg, while substance 2 has a specific heat of 1.0 cal/g-deg at 20°C. The addition of an equivalent amount of heat to 10 grams of either substance will result in substance ____________ increasing more in temperature.

30. In changing a large number into exponential notation form, the decimal point is moved to the ________ by n places and the exponent is equal to ____________.

31. A temperature of 150 K is equal to ____________ on the Fahrenheit scale.

32. A volume of 131 mL of water weighs ____________.

33. A 15.0 mL volume of liquid with a specific gravity of 1.75 weighs ____________.

34. A volume of a clear blue liquid with a density of 0.89 g/cc weighs 43 g. The volume is ________. Its specific gravity is equal to ________.

35. A copper atom weighs 2.3×10^{-25} lb. Its weight in milligrams is ____________, and its corresponding weight in kilograms is ____________.

36. A drop of rain has a volume of about 50 microliters. This is equivalent to ________ quarts (1 quart = 0.946 L).

37. A single bond between two carbon atoms has a length of 1.54 Angstroms (1 Angstrom = 1×10^{-10} m). This bond length is equivalent to ____________ in.

38. A substance is found to have a density of 2.40 g/cc and weighs 100. g. Its volume is ____________.

39. At 20°C, the specific gravity of a solid substance A is 2.46, while the density of a liquid B is 2.0 g/cc. If solid A is placed in liquid B, it will ____________.

40. A block of Blasa wood has dimensions of 2.00 cm x 6.00 cm x 1.50 m and a density of 0.120 g/cc. Its mass is __________.

41. Express the following quantities using the metric (SI) units indicated.

a. 34 cm = __________ m

b. 1×10^3 g = __________ μg

c. 2.3×10^3 g/cc = __________ mg/L

d. 6.3 cal/g-deg = __________ kcal/mg-deg

42. Volumes of liquids may be measured accurately with either a __________, __________ or __________. When only approximate volumes are needed, a __________ is used.

43. The ________ of an object is independent of location, while the ________ of an object is dependent on location.

44. The ______ of a substance is measured on a laboratory balance.

45. The density of a substance usually __________ with increasing temperature because the ________ increases with temperature, while the __________ does not change with temperature.

46. A billion seconds is equivalent to ca. _____ years. A billion seconds ago, most students in this class were probably _____ born yet.

47. Indicate the number of significant figures in the following

a. 2.00 ______________

b. 11,000 ______________

c. 0.0043 ______________

d. 5.32×10^4 ______________

e. 5.0×10^{-3} ______________

48. Carry out the following operations and report the correct number of significant figures in the answer.

a.
```
 14.00
  1.042
307.
```

b.
```
 3,640.64
-   13.1
```

c. 0.37 x 41.56 = ______________

d. $\frac{18.4}{3.468}$ = ______________

e. $\frac{0.1340 \times 6.20}{146.00}$ = ________________

49. When heat is added to a substance, the temperature of the substance increases, with the magnitude of the temperature rise depending on the specific heat of the substance. Distinguish between the terms temperature, heat and specific heat by writing out the definition for each.

ANSWERS FOR FOCUSED REVIEW EXERCISES

Exercise I

1. 1.02×10^{-2}
2. 6.32×10^{-12}
3. 7.00×10^{40}
4. 1.3642×10^{6}
5. 1.1×10^{24}
6. 6×10^{-21}
7. 1.37×10^{-8}
8. 6×10^{-3}
9. 7.423×10^{26}
10. 4.361×10^{30}

Exercise II

1. 1.32×10^{3} mL; 1.32×10^{6} µL; 1.32×10^{9} nL
2. 6.32×10^{-3} g; 6.32×10^{-6} kg; 6.32×10^{-9} Mg
3. 3.9×10^{-4} m; 3.9×10^{-7} km; 3.9×10^{2} µm
4. 3.6×10^{-4} ms; 3.6×10^{-7} s; 3.6×10^{2} ns

Exercise III

1. 12 in; 30 cm; 1.9×10^{-4} miles
2. 10 oz; 0.7 lb; 3×10^{8} µg
3. 1.9×10^{5} µL; 1.9×10^{-1} L; 2.0×10^{-1} qts

ANSWERS TO SELF-TEST QUESTIONS

Multiple Choice

1. a, c, e
2. a (fact); b & d(hypothesis); c(fact)
3. b, d
4. b, d
5. a, c, d
6. a, d
7. b
8. a
9. b
10. b, c, d
11. a, b, c
12. a, b, c
13. b, c
14. b, c
15. a, c
16. b, c
17. b; in this conversion, L must be in the numerator
18. a. one b. seven c. two
    d. ambiguous. Can be two or as many as six.

Completion

1. anything that has mass and also occupies space
2. hypotheses, theories
3. coefficient, exponent
4. a. 1×10^4
   b. 1×10^{-3}
   c. 4.3×10^7
   d. 9.43×10^9
5. 12
6. System International, Metric System
7. meter, m, liter, L
8. 10^5
9. mass
10. time, s
11. identical
12. 573 K
13. 98.6°F
14. 0 K or -273°C, ceases
15. 25 liters
16. solid, liquid, gas
17. heat energy
18. taken away
19. gas > liquid > solid
20. less dense
21. density is a physical quantity and has units, while specific gravity is a ratio of densities and is unitless
22. intensive
23. decreases
24. specific gravity
25. the capacity to do work
26. motion; stored
27. energy can neither be created or destroyed
28. the input of large amounts of energy or heat is necessary to increase the temperature of the substance
29. substance 2
30. left, plus n
31. -189°F
32. 131 g
33. 26.2 g
34. 48 cc (mL), 0.89
35. 1.0×10^{-19} mg, 1.0×10^{-25} kg
36. 5.3×10^{-5} qt
37. 6.06×10^{-9} in
38. 42 cc (mL)
39. sink
40. 216 g
41. a. 0.34 m
   b. $1 \times 10^{+9}$ μg
   c. 2.3×10^9 mg/L
   d. 6.3×10^{-6} cal/mg-deg
42. buret, pipet, volumetric flask, graduated cylinder
43. mass, weight

44. mass
45. decreases, volume, mass
46. 31.7 years. In 1997, a billion seconds ago would be about the year 1965-66. Most students in this class were not born yet.
47. a. 3
    b. at least 2; if all figures are significant, then 5
    c. 2
    d. 3
    e. 2
48. a. 322
    b. 3627.5
    c. 15
    d. 5.31
    e. 0.00569

49. Temperature is the hotness or coldness of a substance, measured by one of the temperature scales. Heat is a form of energy that is measured in calories, with a calorie being the amount of heat that is necessary to raise 1 g of water by 1^{0} C (centigrade). The specific heat is an inherent property of every substance. It is the amount of heat, in calories necessary to raise the temperature of 1 g of the substance by 1 ^{0}C.

Learning in old age is writing on sand but
learning in youth is engraving on stone.
- Arabian proverb

2

Atoms

CHAPTER OBJECTIVES

This chapter begins to present the fundamental aspects of chemistry. Whether or not you have been introduced to most of the terms and concepts previously in high school, you should very carefully master the material since it provides the real starting point and the basic grounding for subsequent chapters. You will also find that an understanding of the biochemistry associated with a living organism relies heavily on the basic concepts and principles of chemistry.

After studying the chapter and working the assigned exercises in the text and the study guide, you should be able to do the following.

1. Summarize and contrast the ideas of Zeno, Democritus, Dalton and Bohr regarding the basic makeup of matter.
2. Name and write out the chemical symbols for the first twenty elements (this is a minimum objective that should be extended).
3. Distinguish between a pure compound and a mixture.
4. State the four basic principles that serve as a basis for Dalton's atomic theory.
5. Characterize the elementary particles in the atom, including their charge, mass and location within the atom.
6. Given the number of protons and neutrons in an atom, determine the charge of the nucleus and the atomic number and mass number for the element.
7. Given the atomic number of an element, locate and identify the element on the periodic table.
8. Distinguish between an atom, isotope, a positive and negative ion and a molecule.
9. Explain the difference between a metal, metalloid, and a nonmetal and locate where each class is located in the Periodic Table.

10. Write out the ground state electronic configuration for at least the first 40 elements, making use of the Periodic Table, the Pauli Exclusion Principle and Hund's Rule.

11. Write out the ground state electronic configuration for positively and negatively charged ions of the first 20 elements.

12. Explain the trend in the ionization energies of the elements in a column (i.e., a group) and in a horizontal row on the periodic table.

13. List the two most abundant elements in the earth's crust and the four most important elements in the human body.

14. Define the important terms and comparisons in this chapter and give specific examples where appropriate.

IMPORTANT TERMS AND COMPARISONS

Atom, Ion and Molecule
Element
Compound and Mixture
Representative Element and Transition Metal
Periodic Table
Dalton's Atomic Theory
Law of Conservation of Mass
Ground State and Excited State
Law of Constant Composition
Shell and Subshell
Unpaired and Paired Electrons
Inner and Outer Shell
Electronic Configuration
Diatomic and Polyatomic Elements
Pauli Exclusion Principle
Atomic Weight
Halogen
Group Numbers and Group Names
Metal, Metalloid and Nonmetal
Isotopes
Noble Gases
Quantized Energy Levels
Ionization Energy
s, p, and d Orbitals
Proton, Neutron and Electron
AMU
Nucleons and Nucleus
Mass Number and Atomic Number Monatomic,
Hund's Rule
Lewis Structure

SELF-TEST QUESTIONS

MULTIPLE CHOICE. In the following exercises, select the correct answer from the choices listed. In some cases, two or more answers will be correct.

1. The concept of quantized energy levels in the atom was first proposed by:

    a. Zeno
    b. Bohr
    c. Lavoisier
    d. Democritus

2. Dalton's atomic theory includes the following statements:

    a. All atoms of the same element are identical to each other.
    b. All atoms are colorless.
    c. Atoms combine to form molecules. The molecule is a tightly bound group of atoms that acts as a unit.
    d. All matter is made up of very tiny indivisible particles called atoms.

3. The mass number of an atom that has 20 neutrons and a charge of +18 in the nucleus is:

a. 20
b. 38
c. 56
d. cannot determine

4. The atomic number of the atom in question 3 is:

a. 20
b. 38
c. 18
d. cannot determine

5. The ground state electronic configuration for the atom which has 6 protons and 6 neutrons in the nucleus is:

	1s	2s	$2p_x$	$2p_y$	$2p_z$	3s
a.	↓↑	↓↑	↓↑			
b.	↓↑	↓↑	↓	↓		
c.	↓↑	↓↑	↓↑	↓↑	↓↑	↓↑
d.	↓↑	↓↑	↓			↓

6. A possible excited state electronic configuration for the atom in question 5 is answer ________________.

7. Which of the following elements is a metalloid?

a. Hydrogen, H
b. Germanium, Ge
c. Silver, Ag
d. Helium, He

8. Which of the following collection of elements are nonmetals?

a. Na, K, Rb
b. F, Cl, Br
c. O, S, Se
d. Cu, Ag, Au

9. The four most abundant elements in the earth's crust, listed in increasing order, are:

a. Ca < Fe < O < Si
b. Fe < Al < Si < O
c. N > C > H > O
d. Cl < O < N < P

10. The mass of a single atom of lead (Pb) is:

a. 3.5 g
b. 3.5 ng
c. 3.5×10^{-22} g
d. 3.5×10^{6} g

11. The four most abundant elements in the human body are:

a. H, Li, C & N
b. C, N, O & H
c. C, N, S, & H
d. Fe, C, N & Na

12. The four most common elements in the human body are considered:

a. transition metals
b. metalloids
c. representative elements
d. metals

13. The two most abundant elements in the crust of the earth are:

a. Ag & Au
b. O & Si
c. H & C
d. H & He

14. An example of periodicity can be seen in comparisns of:

a. H, Li, Na & K
b. C, N, Al & Si
c. H, He, B & C
d. Li, Be, C & N

15. The Lewis structure for sulfur is:

a. $:\ddot{S}:$
b. $:S$
c. $:S:$
d. $:\underset{\cdot\cdot}{\ddot{S}}:$

<u>COMPLETION.</u> Write the word, phase or number in the blank which will complete the statement or answer the question.

1. The three elementary particles in an atom are the __________, __________ and __________, with only __________ of them being classified as nucleons.

2. Of the alkali metals, Li, Na, K, Rb and Cs, ____________ requires the least energy to ionize an electron off and to produce a positive ion.

3. The chemical symbols for the elements beryllium, arsenic, potassium and sodium are ________, ________, ___________ and __________, respectively.

4. The atomic nucleus always possesses a __________ charge because the __________ resides in the nucleus.

5. Isotopes of an element have the same number of __________, but different number of __________. Therefore, their __________ numbers are the same although their __________ numbers differ.

6. A positive ion is produced from an atom by __________ an __________ the atom.

7. The halogens are in Group __________ and have __________ electrons in their outer shell.

8. Calcium, with atomic number 20, is in Group IIA. It is an example of a __________ element which has __________ electrons in its outer shell.

9. The ionization energy of elements in Group IA __________ going down the column, while the ionization energy going across a horizonal row of elements (from left to right) generally becomes __________.

10. The periodic table works because elements in the same __________ have the same __________ in the __________ shell.

11. In going down Group VIIA from flourine to astatine, the molecular weights progressively increase. Generally, as the molecular weights of similar compounds increase, the boiling points and melting points __________________ .

ANSWERS TO SELF-TEST QUESTIONS

Multiple Choice

1. b
2. a, c, d
3. b
4. c
5. b
6. a or d
7. b
8. b, c
9. b
10. c
11. b
12. c
13. b
14. a
15. a

Completion

1. proton, neutron, electron, 2
2. Cesium (Cs)
3. Be, As, K and Na
4. positive, proton
5. protons and electrons, neutrons, atomic, mass
6. pulling off or ionizing, electron from
7. VII, 7
8. representative or a metallic, 2
9. decreases, greater
10. group or column, number of electrons, outer
11. increase

"I believe the chemical bond is not so simple as some people seem to think."
- Robert S. Mulliken

3

Chemical Bonds

CHAPTER OBJECTIVES

Although much remains to be learned about the atom, the titles of this and all subsequent chapters reveal that a great deal of attention is devoted to understanding the chemistry of ions and molecules. In either case, the outer **electronic configuration of the atom is altered** by the gain or loss of an electron to form a charged particle, or alternatively, electrons are shared in a common covalent bond between two atoms. As a result, the chemical and physical properties of these chemical species are different than the atoms from which they were derived. Also, in the case of complex ions or molecules, the species take on specific three-dimensional shapes, unlike that of the beautifully simple spherical shape of an atom. The molecular shapes can be predicted from the **valence-shell electron-pair repulsion (VSEPR)** theory.

After you have studied this chapter and worked the exercises in the text and study guide, you should be able to do the following.

1. Write the equations for the formation of a cation or anion from the parent atom.
2. Using the octet rule as a guide, predict why certain ions are stable, while others are nonexistent.
3. From a knowledge of the charge on the positive and negative ions and using the principle of electrical neutrality, predict the formulas for a variety of ionic compounds.
4. Distinguish between an atom, an ion and a molecule.
5. Describe the characteristic difference between an ionic bond and a covalent bond.
6. Characterize a single-, double- and triple-bond and give specific examples of compounds containing these bond types.
7. Write out the molecular formula for a variety of compounds, in addition to drawing the structural formula and Lewis structure for each compound.

8. State the VSEPR Theory. Given the number of valence shell electron pairs of the central atom of a molecule and the nature of bonded atoms, predict the shape of the molecule and the size of the angle between the bonds in each. Give specific examples of molecules with each shape.

9. Define electronegativity and explain the trend in electronegativities in the periodic table as shown in Table 3.5 (text).

10. Draw the structural formulas for some di-, tri- and tetraatomic molecules. Given the electronegativity value for each atom, indicate whether the individual **bonds** and whether the **molecules** themselves are polar or nonpolar.

11. Using the general rules which relate the electronegativity differences of the atoms in a bond to the bond type, predict whether the bonds in a variety of diatomic molecules are ionic, polar-covalent or nonpolar-covalent.

12. Write out the name, together with the molecular and structural formulas for some common compounds (utilize Table 3.2 & 3.4, text) and some medicinally important polyatomic ions (Box 3B, text).

13. Write out the names and chemical formulas for some of the inorganic compounds that are found in the chapter.

14. Define the important terms and comparisons in this chapter and give specific examples where appropriate.

IMPORTANT TERMS AND COMPARISONS

Ion, Cation, Anion
Ionic and Covalent Bonds
Three-Dimensional Molecular Structure
Polyatomic Ions
Bond Angle
VSEPR Theory
Shared and Unshared Electron Pair
Single, Double and Triple Bond
Electronegativity
Molecular and Structural Formula
Lewis Structure
Electrical Neutrality
Hemoglobin
Linear and Angular Molecules
Trigonal, Pyramidal and Tetrahedral Shapes
Partial Charges, $\delta\pm$
Polar and Non-Polar Bonds
Polar and Non-Polar Molecules

SELF-TEST QUESTIONS

MULTIPLE CHOICE. In the following exercises, select the correct answer from the choices listed. In some cases, two or more answers will be correct.

1. The formula for cesium phosphate is:

a. $Cs_3(PO_4)$
b. $Cs_2(P_4)$
c. $Cs(PO_4)_3$
d. $CsPO_4$

2. The Lewis structure for the ozone molecule, O_3, is:

a. 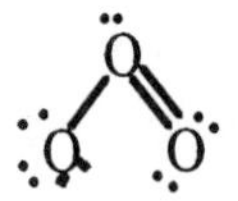

c. 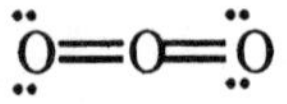

b. 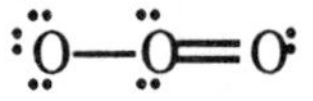

d. 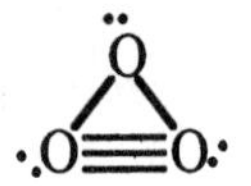

3. Draw the Lewis structure for H2CO. Indicate the number of electrons and the types of electron pairs that are around the oxygen atom.

a. 3 bonding pairs and 1 non-bonding pairs
b. 2 bonding pairs and 2 non-bonding pairs
c. 4 non-bonding pairs
d. 4 bonding pairs

4. Which of the following ions is stable?

a. F^+
b. S^{2-}
c. Ca^+
d. O^{2+}
e. I^-
f. N^{3-}

5. A particular atom has the following characteristics. The atomic number is 12 and it forms cations with a +2 charge. This element is:

a. Na
b. Al
c. S
d. Mg

6. The bond that is most polar is:

a. O-H
b. S-H
c. B-H
d. F-H

7. The order of decreasing difference in polarity in the bonds in question 6 is:

a. H-F > O-H > S-H > B-H
b. H-F > S-H > B-H > O-H
c. O-H > H-F > S-H > B-H
d. S-H > O-H > H-F > B-H

8. Which of the following molecules have a correct Lewis structure?

a. 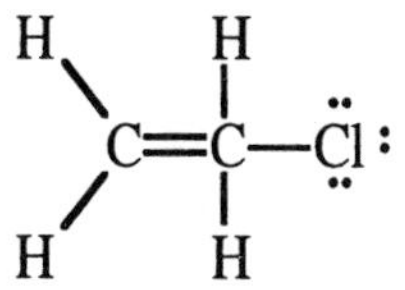

c. [:C≡N:]$^{-1}$

b. 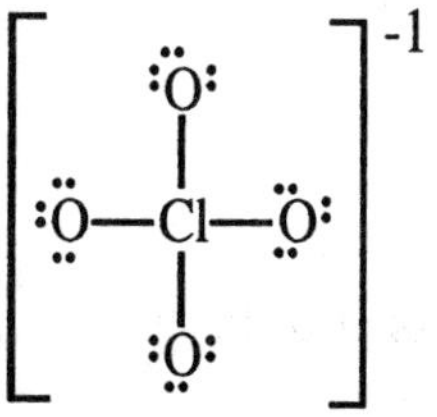

d. :C≡O:

9. Indicate which of the following compounds are covalent.

a. $CaCl_2$
b. ClBr
c. SO_2
d. KCl

10. Indicate which pair of atoms has the greatest electronegativity difference.

a. H & C
b. H & O
c. F & Cl
d. C & N

11. Predict the molecular formula for the compound that contains calcium and phosphate.

a. $Ca(PO_4)$
b. $Ca_3(PO_4)_2$
c. $Ca_2(PO_4)_3$
d. $Ca(PO_4)_2$

12. Indicate which of the following pairs of compounds is not named correctly.

a.	CuCl & $CuCl_2$	cuprous chloride and cupric chloride
b.	$FeBr_3$ and $FeBr_2$	iron tribromide & iron dibromide
c.	SnO and SnO_2	stannous oxide & stannic oxide
d.	NO and N_2O	nitrogen oxide & dinitrogen oxide

13. Indicate which of the following statements about the relative electronegativity of two atoms is not correct.

a. Br is greater than As.
b. Li is greater than K
c. B is less than Al
d. Na is less than Mg

<u>COMPLETION.</u> Write the word, phrase or number in the blank that will complete the statement or answer the question.

1. Based on the VSEPR theory, determine the molecular shape of each molecule.

a. AlF_4^- ____________________

b. C_2HCl ____________________

c. H_3O^+ ____________________

d. PCl_3 ____________________

e. BI_3 ____________________

f. I_3^- ____________________
(The central I does not obey the octet rule.)

2. Classify whether the bond between the two elements shown is non-polar, covalent, polar covalent or ionic.

a. C, C ____________________

b. Li, F ____________________

c. Na, O ____________________

d. H, I ____________________

e. O, S ____________________

3. Name these compounds or ions.

a. CN^- ____________________

b. $(NH_4)_2SO_4$ ____________________

c. NO_2^- ____________________

d. K_2O ____________________

e. $MgCl_2$ ____________________

f. $Li(CH_3COO)$ ____________________

4. Indicate the bond angles in these molecules.

a. CO_2 ____________________

b. CCl_4 ____________________

c. H_2O ____________________

d. BCl_3 ____________________

e. C_2H_2 ____________________

5. Write out the chemical formulas for these compounds.

a. sodium bicarbonate ____________________

b. rubidium sulfate ____________________

c. magnesium permanganate ____________________

d. ammonium oxide ____________________

6. The formation of a positively charged __________ from an atom requires the __________ of an electron, while the formation of a negatively charged __________ requires the __________ of an electron.

7. A __________ bond involves the __________ of one or more electron pairs, while an __________ bond results from an electrostatic attraction between __________ particles.

8. Although the C^{4+} cation has a complete outer shell, it does not exist because __________ are unstable.

9. In solid NaCl, each Na^+ has __________ Cl^- ions as ______________________________ ______________________________.

10. An __________ orbital has a spherical shape, while a p orbital is shaped like a ______________.

11. Name these compounds.

a. $NiCl_2$ ____________________

b. $Cr(NO_3)_3$ ____________________

c. $NaH(SO_4)$ ____________________

d. MgF_2 ____________________

12. Write the chemical formula for each of these compounds.

a. copper(II) iodide ____________________

b. sodium oxide ____________________

c. aluminum oxide ____________________

13. In the formation of the F_2 molecule, the p_z atomic orbital from each of the two fluorine atoms overlap each other to form a molecular orbital . This molecular orbital, called a ____________ orbital, contains ____________ shared electrons which manifest the covalent bond in the molecule.

ANSWERS TO SELF-TEST QUESTIONS

Multiple Choice

1. a
2. a (note that 18 valence electrons are involved and the octet rule must be obeyed)
3. b
4. b, e, f
5. d
6. d
7. a
8. b, c, d
9. b, c
10. b
11. c
12. b
13. b

Completion

1. a. tetrahedral
   b. linear
   c. pyramidal
   d. pyramidal
   e. trigonal planar
   f. linear

2. a. nonpolar covalent, $\Delta E = 0$
   b. ionic, $\Delta E = 3.0$
   c. ionic, $\Delta E = 2.4$
   d. nonpolar covalent, $\Delta E = 0.4$
   e. polar covalent, $\Delta E = 1.0$

3. a. cyanide anion (or ion)
   b. ammonium sulfate
   c. nitri_t_e anion (or ion)
   d. potassium oxide
   e. magnesium chloride
   f. lithium acetate

4. a. 180°
   b. 109.5°
   c. 105°
   d. 120°
   e. 180°

5. a. $Na(HCO_3)$
   b. $Rb_2(SO_4)$
   c. $Mg(MnO_4)_2$
   d. $(NH_4)_2O$

6. cation, removal, anion, addition
7. covalent, sharing, ionic, charged
8. concentrated charges on one atom
9. 6, nearest neighbors
10. s, dumbbell

11. a. nickel(II) chloride
    b. chromium(III) nitrate
    c. sodium hydrogen sulfate
    d. magnesium fluoride

12. a. CuI_2
b. Na_2O
c. Al_2O_3

13. sigma, 2

....chemical transformations....are
responsible for corrosion and decay,
for development, growth and life.
- C.N. Hinshelwood
The Kinetics of Chemical Change, 1940

4

Chemical Reactions

CHAPTER OBJECTIVES

Chemistry is a quantitative science. This means that it is possible to determine the exact quantity of a substance - for example, the volume, density, mass or number of molecules of water in a raindrop. Importantly, one can relate **macroscopic** quantities, such as the mass of a substance, to **microscopic** quantities, such as the number of atoms or molecules in a given quantity of the substance. In addition, exact relationships can be represented by a balanced chemical equation or a mathematical equation. For example, we all realize that it requires gasoline to run an automobile. This process of combustion of the gasoline can be described quantitatively by a balanced chemical equation. This chapter presents these concepts, relationships and procedures that provide the basis for some of the quantitative aspects of chemistry.

After you have studied the chapter and worked the exercises in the text and study guide, you should be able to do the following.

1. Calculate the formula weight of any ionic or molecular compound from its molecular formula.

2. Calculate the number of moles in a given mass of a compound.

3. Using Avogadro's number of 6.02×10^{23}, determine the number of molecules in a given mass of a compound.

4. Readily interconvert between the mass of a substance and the number of moles and molecules of the compound.

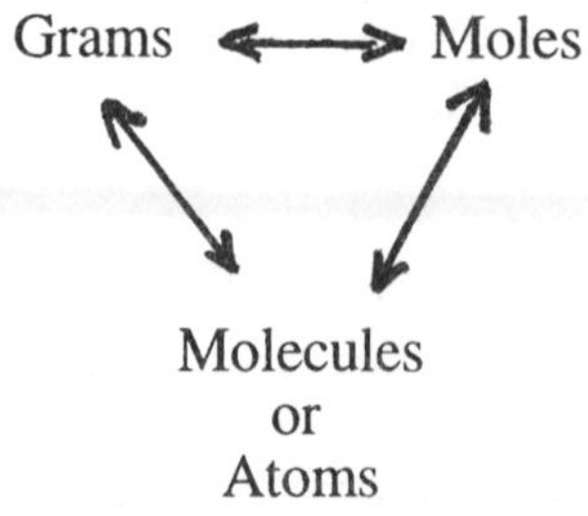

5. Using the atomic weights, determine the relative mass of pairs of atoms on the periodic table.

6. Balance simple chemical equations.

7. From a balanced chemical equation, carry out calculations to determine the mass and molar relationship between reactants and products.

8. Using the balanced equation and given the actual quantity of reactants used up and of the product formed in a reaction, calculate the theoretical and percentage yield of product.

9. Write out the complete, balanced equations for the reaction of various ionic compounds in aqueous solution. Point out the spectator ions and then rewrite the ionic equation.

10. Write out the general reaction scheme for a combination and a decomposition reaction and then give a specific example of each.

11. Be able to recognize and balance a redox reaction. Indicate the species that are oxidized and reduced and the species which are the oxidizing and the reducing agent, respectively.

12. Explain how a voltaic cell works.

13. Given balanced equations, including the heat gain or loss, generally classify the reaction type and indicate whether the reaction is exothermic or endothermic.

14. Define the important terms and comparisons in this chapter and give specific examples where appropriate.

IMPORTANT TERMS AND COMPARISONS

Chemical Reaction
Reactant and Product
Formula Weight and Molecular Weight
AMU
Avogadro's Number and Mole
Factor-Label Method
Conversion Factor
Balanced Equation
Coefficient and Subscript
Conservation of Mass
Actual Yield and Theoretical Yield
Side Reaction
Aqueous
Combination and Decomposition Reactions
Voltaic Cell
Symbols, $\uparrow$, $\downarrow$
Oxidation and Reduction
Oxidizing and Reducing Agent
Combustion
Limiting Reagent
Respiration
Rusting
Metabolism
Heat of Reaction and Heat of Combustion
Exothermic and Endothermic
Exergonic and Endergonic
Calorie
Heat of Combustion
Antiseptic and Disinfectant

FOCUSED REVIEW

To become comfortable and efficient with the mathematical manipulations of the new material, work the following exercises before going onto the SELF-TEST QUESTIONS.

Exercise I. Calculation of Formula Weights. From the molecular formulas, calculate the formula weight of each compound. Round the answer off to the first decimal place.

a. $Mg(ClO_4)_2$ ____________________

b. Na_2O ____________________

c. $Al(H_3CCO_2)_3$ ____________________

d. $FeBr_3$ ____________________

e. O_3 ____________________

f. $C_{18}H_{36}O_2$ ____________________

g. $C_3H_7NO_2$ ____________________

h. $Pt(NH_3)_2Cl_2$ ____________________

<u>Exercise II.</u> <u>Interconversion of Grams, Moles, and Molecules</u>. The most important interrelationship in this chapter is that between grams, moles and molecules and is represented below.

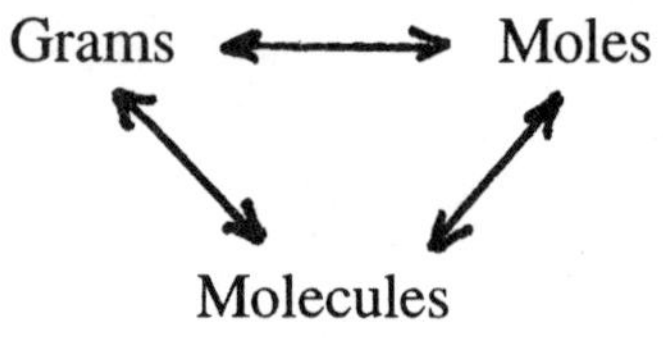

In converting one of these quantities to another, a conversion unit is available and is used "as is" or inverted, depending on the direction of the conversion (i.e., whether converting from grams to moles or from moles to grams).

Using H_2O as an example, it will be (or is) obvious that 1 gram of water is equivalent to the following.

$$1.0 \text{ g } H_2O \leftrightarrow 5.5\text{x}10^{-2} \text{ moles } H_2O \leftrightarrow 3.3\text{x}10^{22} \text{ molecules } H_2O$$

To accomplish these interconversions, the quantity is multiplied by the appropriate conversion factor, making use of the factor-label method. The conversion factor in each of these examples is underlined.

For example:

Given: 1.0 g H_2O Convert to: moles H_2O

$$\left(\frac{1.0 \cancel{\text{g H}_2\text{O}}}{1}\right)\underline{\left(\frac{1 \text{ mole H}_2\text{O}}{18 \cancel{\text{g H}_2\text{O}}}\right)} = 5.5\text{x}10^{-2} \text{ moles H}_2\text{O}$$

Given: 1.0 g H_2O Convert to: molecules H_2O

$$\left(\frac{1.0 \cancel{\text{g H}_2\text{O}}}{1}\right)\left(\frac{1 \cancel{\text{mole}} \text{ H}_2\text{O}}{18 \cancel{\text{g H}_2\text{O}}}\right)\underline{\left(\frac{6.02\text{x}10^{23} \text{ molecules}}{1 \cancel{\text{mole}} \text{ H}_2\text{O}}\right)} = 3.3\text{x}10^{22} \text{ molecules H}_2\text{O}$$

Given: $5.5x10^{-2}$ moles H_2O Convert to: g H_2O

$$\left(\frac{5.5x10^{-2} \text{ moles } H_2O}{1}\right)\left(\frac{18 \text{ g}}{1 \text{ mole } H_2O}\right) = 1.0 \text{ g } H_2O$$

Given: $5.5x10^{-2}$ moles H_2O Convert to: molecules H_2O

$$\left(\frac{5.5x10^{-2} \text{ moles } H_2O}{1}\right)\left(\frac{6.02x10^{23} \text{ molecules}}{1 \text{ mole } H_2O}\right) = 3.3x10^{22} \text{ molecules } H_2O$$

Given: $3.3x10^{22}$ molecules H_2O Convert to: moles H_2O

$$\left(\frac{3.3x10^{22} \text{ molecules } H_2O}{1}\right)\left(\frac{1 \text{ mole } H_2O}{6.02 \text{ x } 10^{23} \text{ molecules}}\right) = 5.5x10^{-2} \text{ moles } H_2O$$

Given: $3.3x10^{22}$ molecules H_2O Convert to: g H_2O

$$\left(\frac{3.3x10^{22} \text{ molecules } H_2O}{1}\right)\left(\frac{1 \text{ mole } H_2O}{6.02x10^{23} \text{ molecules}}\right)\left(\frac{18 \text{ g } H_2O}{1 \text{ mole}}\right) = 1.0 \text{ g } H_2O$$

As noted in the text, the conversion from grams to moles or visa versa requires only one step, while conversion between grams and molecules requires 2 steps or 2 conversion factors. The importance of using the factor-label method is brought out in these calculations as well.

Carry out the following conversions, indicating the conversion unit used in the factor-label analysis. In addition, the answer should have the correct number of significant figures.

A. GRAMS ↔ MOLES CONVERSION

	Grams	Conversion Factor	Moles
(i)	4.0 g HF	____________	____________
(ii)	33 g KOH	____________	____________
(iii)	0.020 g AlF_3	____________	____________
(iv)	____________	____________	0.30 moles PCl_3
(v)	____________	____________	1.2 moles GeH_4
(vi)	____________	____________	0.01 moles H_2CO_2

B. MOLES ↔ MOLECULES CONVERSION

	Grams	Conversion Factor	Molecules
(i)	0.34 moles H_2	______	______
(ii)	11.3 moles Na_2SO_4	______	______
(iii)	$3.4x10^{-24}$ moles NH_3	______	______
(iv)	______	______	1 molecule of I_2
(v)	______	______	$6x10^{20}$ molecules O_3
(vi)	______	______	$4x10^{6}$ molecules $HClO_2$

C. GRAMS ↔ MOLECULES CONVERSION

	Grams	Conversion Factor	Molecules
(i)	0.14 g NCl_3	______	______
(ii)	4.0 mg S_8	______	______
(iii)	7.4 ug HI	______	______
(iv)	______	______	$4x10^{27}$ molecules $SeCl_2$
(v)	______	______	$6x10^{12}$ molecules $C_2H_4O_2$
(vi)	______	______	$2x10^{2}$ molecules GeI_4

Exercise III. Stoichiometry

In determining the weight relationships in chemical reactions, the key points are these:

1. The equation must be **balanced.**

2. **Molecules** react with **molecules**. **Moles** react with **moles**. The coefficients in the balanced equation dictate exactly how many of each of the reactants will react to form exactly how many product molecules. Therefore, the number of **moles** of the known reactant or product must be calculated first, if it is not given.

3. The coefficients in the balanced equation indicate the relative number of moles of each reactant and product. Therefore, multiply the moles of the known compound by the coefficient of the compound whose weight is unknown divided by the coefficient of the compound of known weight (or moles).

4. Convert the number of moles back to weight, by use of the appropriate conversion factor.

Consider the following equation. From the quantity given, determine the number of moles of the other quantities of interest.

C_7H_{16} + 11 O_2 → 7 CO_2 + 8 H_2O

A. Given: 0.1 mole C_7H_{16}

moles O_2	moles CO_2	moles H_2O
________	________	________

B. Given: 0.1 mole O_2

moles C_7H_{16}	moles CO_2	moles H_2O
________	________	________

C. Given 0.10 mole CO_2 produced

moles C_7H_{16}	moles O_2	moles H_2O
________	________	________

Exercise IV. Limiting Reagent. In most "real-life" situations, there are reactions or relationships that involve a limitation or a restriction that there is a limited amount of one of the reactants and therefore, the amount of product formed will be limited.

A few simple (non-chemistry) examples may help.

A football program has all the football gear (shoulder pads, helmets, etc) to have a team of100 players. However, they have only 75 shirts with the player's numbers on them. As a result, they can have only 75 players on the team. The number of shirts is limiting and the shirt is the limiting "reagent".

A chemistry course is taught in a classroom that can accommodate 150 students. However, the required laboratory can only accommodate 50 students. The enrollment in the course is limited to 50. The limiting "reagent" is the number of students that can be accommodated in the laboratory.

Let's try a chemical example.

	1 Na	+ 1/2 Cl_2	= 1 NaCl
	22.9 g	200 g	
Available reagents	1 mole	2.82 moles	-----
Used reagent	1 mole	0.5 moles	-----
Final products of the reaction	-----	2.32 moles	1 mole

An experimenter carries out the reaction with 22.9 g of Na metal and 200 g of Cl_2 gas. Calculations indicate (verify this) that there is 1 mole of Na and 2.82 moles of chlorine gas. The balanced equation indicates that 1 mole of Na will react with 1/2 mole of Cl_2 gas to produce 1 mole of NaCl. Therefore, there is an excess of Cl_2 (2.82 - 0.50 = 2.32 moles) that will go unreacted because there will not be enough Na to react with it – that is, Na is the limiting reagent in these reaction conditions. The extent of the reaction is limited by the Na that is available. At the end of the reaction, all the Na will be used up, one mole of NaCl will be produced and 2.32 moles of Cl_2 will remain unreacted.

SELF-TEST QUESTIONS

MULTIPLE CHOICE. In the following exercises, select the correct answer from the choices listed. In some cases, two or more answers will be correct.

1. The reaction of C_4H_{10} with O_2 to produce CO_2 and H_2O is what type of reaction?

   a. Precipitation
   b. Combustion
   c. Rusting
   d. Respiration

2. The formula weight for $Ca_3(PO_4)_2$ is:

   a. 279.3
   b. 310.3
   c. 167.3
   d. 543.3

3. A thimble of water contains 4.0×10^{21} molecules. The mass of water is:

   a. 1.1 g
   b. 11 g
   c. 1.1×10^{-8} g
   d. 0.11 g

4. Calculate the number of moles of aspirin, $C_9H_8O_4$, in a 4.0 g tablet.

   a. 2.3×10^{-4}
   b. 2.3
   c. 4.6×10^{-3}
   d. 0.023

5. The number of grams of K_2O produced from the oxidation of 4.0 grams of K is:

   a. 4.7 g
   b. 4.0 g
   c. 9.4 g
   d. 6.3 g

6. If the reaction in problem 5 actually produces only 1.0 g K_2O, what is the percentage yield for the reaction?

   a. 25%
   b. 11%
   c. 21%
   d. 16%

7. In the equation below, the oxidizing agent in the reaction is:

   $$Al_{(s)} + H^+ = Al^{+3} + H_{2(g)}$$

   a. $Al_{(s)}$
   b. H^+
   c. Al^{+3}
   d. H_2

8. The species which is oxidized in problem 7 is:

   a. $Al_{(s)}$
   b. H^+
   c. Al^{+3}
   d. H_2

9. Indicate which reaction is the most endothermic.

   a. $A + B \rightarrow C + 1{,}600$ cal.
   b. $D + E + 12{,}000$ cal. $\rightarrow F$
   c. $G + H \rightarrow I + 2.3$ kcals.
   d. $X + Y + 2.6$ kcal. $\rightarrow Z$

10. A chemist used a laboratory calorimeter to determine that the combustion of glucose produced 686 kcal per mole of glucose. The amount of energy produced in the body by the metabolism of this same amount of glucose is:

   a. Greater than 686 kcal.
   b. Less than 686 kcal.
   c. Equal to 686 kcal.
   d. Can not predict

11. The correct answer to problem 10 derives from a law which states that the heat of reaction depends on the starting compounds and the final products and is independent of the reaction pathway. This law is called:

   a. Law of Balanced Reactions
   b. Avogadro's Law
   c. Hess's Law
   d. Hund's Rule

COMPLETION. Write the word, phrase or number in the blank which will complete the statement or answer the question.

1. The mass of germanium, Ge, (atomic number 32 and atomic weight 72.59) which contains a mole of Ge atoms is __________.

2. A __________ is used to determine the __________ value of foods.

3. The __________ ______ __________ is measured in a calorimeter. This is determined by measuring the temperature change of the water in the calorimeter. If the temperature decreases as a result of the reaction, the reaction is __________.

4. An __________ reaction is one in which __________ is given off, but not necessarily as __________.

5. __________ in organic compounds is defined as the gain of hydrogen or the loss of __________.

6. Four very common reaction types in aqueous solution which involve ions are (1) __________, (2) __________, (3) __________ or (4) __________.

7. Complete the following table.

			------- Number of -------	
	Compound	Grams	Moles	Molecules
a.	NH_3	__________	$6.0\text{x}10^{-4}$	__________
b.	$C_2H_2Br_2$	6.3	__________	__________

c. $Si_2C_6H_{18}$ __________ __________ $9.0x10^{12}$

8. Complete and balance the following equations.

a. $C_5H_{12} + O_2 \rightarrow CO_2 + H_2O$

b. $Al^{+3} + SO_4^{2-} \rightarrow$

c. $SnO_2 + H_2 = Sn_{(s)} + H_2O$

d. $Au + Cl_2 \rightarrow AuCl_3$

e. $C_6H_{12}N_4 + H_2O \rightarrow H_2CO + NH_3$

f. $HBr + KOH \dashrightarrow KBr + H_2O$

9. Indicate the type of reaction that is written out in question 8, a-d and f.

10. Consider the reaction of X and Y to form Z, with the indicated stoichiometry.

$$X + 2Y \rightarrow Z$$

The formula weights are: X = 50; Y = 200; Z = 100

a. The reaction indicates that __________ moles of X react with __________ moles of Y to produce __________ moles of Z.

b. If 200 grams of X is consumed in the reaction, the number of grams of Y consumed is __________ and the number of moles of Z produced is __________.

11. Consider the same stoichiometric equation in question 9. If 3 moles of Z were formed in the reaction, then __________ grams of X reacted with __________ grams of Y.

12. Consider the reaction of magnesium metal with oxygen gas.

$$Mg + O2 = MgO$$

The reaction is carried out using 24.3 g of Mg and 32 g of O2. The limiting reagent, which will be completely used up, is ________ . The reaction with produce __________g of MgO and leave an excess of ______g of the unreacted _____.

ANSWERS FOR FOCUSED REVIEW EXERCISES

Exercise I.

a. 223.2
b. 62.0
c. 204.0
d. 295.5
e. 48.0
f. 284.0
g. 89.0
h. 300.0

Exercise II.

A. (i) $\frac{1 \text{ mole HF}}{20 \text{ g HF}}$; 0.20 moles

(ii) $\frac{1 \text{ mole KOH}}{56.1 \text{ g KOH}}$; 0.59 moles

(iii) $\frac{1 \text{ mole } AlF_3}{84 \text{ g } AlF_3}$; 2.4×10^{-4} moles

(iv) $\frac{137.4 \text{ g } PCl_3}{1 \text{ mole } PCl_3}$; 41 g PCl_3

(v) $\frac{76.6 \text{ g } GeH_4}{1 \text{ mole } GeH_4}$; 92 g GeH_4

(vi) $\frac{46 \text{ g } H_2CO_2}{1 \text{ mole } H_2CO_2}$; 0.46 g H_2CO_2

B. (i) $\frac{6.02 \times 10^{23} \text{ molecules}}{1 \text{ mole } H_2}$; 2.0×10^{23} molecules

(ii) $\frac{6.02 \times 10^{23} \text{ molecules}}{1 \text{ mole } Na_2SO_4}$; 6.8×10^{24} molecules

(iii) $\frac{6.02 \times 10^{23} \text{ molecules}}{1 \text{ mole } NH_3}$; 2.0 molecules

(iv) $\frac{1 \text{ mole } I_2}{6.02 \times 10^{23} \text{ molecules}}$; 4.2×10^{-22} g.

(v) $\frac{1 \text{ mole } O_3}{6.02 \times 10^{23} \text{ molecules}}$; 0.48 g.

(vi) $\frac{1 \text{ mole } HCl0_2}{6.02 \times 10^{23} \text{ molecules}}$; 4.6×10^{-16} g.

C. (i) $\left(\frac{1 \text{ mole}}{120.4 \text{ g } NCl_3}\right)\left(\frac{6.02 \times 10^{23} \text{ molecules}}{1 \text{ mole}}\right)$; 7×10^{20} molecules

(ii) $\left(\frac{1 \text{ mole}}{256.6 \text{ g } S_8}\right)\left(\frac{6.02 \times 10^{23} \text{ molecules}}{1 \text{ mole}}\right)$; 9×10^{18} molecules

(iii) $\left(\frac{1 \text{ mole}}{127.9 \text{ g HI}}\right)\left(\frac{6.02 \times 10^{23} \text{ molecules}}{1 \text{ mole}}\right)$; 3×10^{16} molecules

(iv) $\left(\frac{149.9 \text{ g } SeCl_2}{1 \text{ mole}}\right)\left(\frac{1 \text{ mole}}{6.02 \times 10^{23} \text{ molecules}}\right)$; 9.9×10^{5} g

(v) $\left(\frac{60 \text{ g } C_2H_4O_2}{1 \text{ mole}}\right)\left(\frac{1 \text{ mole}}{6.02\text{x}10^{23} \text{ molecules}}\right)$; $6\text{x}10^{-10}$ g

(vi) $\left(\frac{580.02 \text{ g } GeI_4}{1 \text{ mole}}\right)\left(\frac{1 \text{ mole}}{6.02\text{x}10^{23} \text{ molecules}}\right)$; $1.9\text{x}10^{-19}$ g

Exercise III.

A. 1.1 moles O_2; 0.7 moles CO_2; 0.8 moles H_2O

B. $9\text{x}10^{-3}$ moles C_7H_{16}; $6\text{x}10^{-2}$ moles CO_2; $7\text{x}10^{-2}$ moles H_2O

C. 0.01 moles C_7H_{16}; 0.16 moles O_2; 0.11 moles H_2O

ANSWERS TO SELF-TEST QUESTIONS

Multiple Choice

1.	b	5.	a	9.	b
2.	b	6.	c	10.	c
3.	d	7.	b	11.	c
4.	d	8.	a		

Completion

1. 72.59 g
2. calorimeter, caloric value
3. heat of reaction, endothermic
4. exergonic, energy, heat
5. reduction, oxygen
6. precipitation, gas formation, acid neutralization of a base, redox
7. a. 0.010 g, $3.6\text{x}10^{20}$ molecules

   b. $3.4\text{x}10^{-2}$ moles, $2.0\text{x}10^{22}$ molecules

   c. 2.2 ng, $1.5\text{x}10^{-11}$ moles
8. a. $C_5H_{12} + 8O_2 \rightarrow 5CO_2 + 6H_2O$

   b. $2Al^{+3} + 3SO_4^{-2} \rightarrow Al_2(SO_4)_3$

   c. $SnO_2 + 2H_2 = Sn_{(s)} + 2H_2O$

   d. $Au + 3/2Cl_2 = AuCl_3$

   e. $C_6H_{12}N_4 + 6H_2O = 6H_2CO + 4NH_3$
9. a. combustion
   b. combination

c. single displacement
d. combination
f. double displacement

10. a. one, two, one
    b. 1600 g, 4 moles

11. 150 g, 1200 g

12. Mg, 40.3 g, 16 g, O_2

Happy is he who gets to know the
reasons for things.
- Motto of Churchill College, Cambridge

5

Gases, Liquids and Solids

CHAPTER OBJECTIVES

After studying the chapter and working the assigned exercises in the text and study guide, you should be able to do the following.

1. State the basic assumptions in the kinetic-molecular theory of gases. Indicate how each individual assumption helps in understanding the behavior or properties of ideal gases.

2. Explain how a barometer and a manometer are setup to read gas pressure.

3. Name the four properties of a gas which defines its physical state and indicate the common units for each.

4. Apply Boyle's Law ($P_1V_1 = P_2V_2$, conditions of constant temperature) to calculate the final volume or pressure of a gas.

5. Use Charles's Law ($V_1/T_1 = V_2/T_2$, conditions of constant pressure) to calculate the final volume or temperature of a gas.

6. Utilize Gay-Lussac's Law ($P_1/T_1 = P_2/T_2$, conditions of constant volume) to calculate the final temperature or pressure of a gas.

7. Apply the combined gas law ($P_1V_1/T_1 = P_2V_2/T_2$) to calculate the final pressure, temperature or volume of a gas.

8. Use the ideal gas law ($PV = nRT$) to calculate either the pressure, volume, the number of moles or the temperature of the gas.

9. Utilize the modified form of the ideal gas law to calculate the molecular weight or the mass of a quantity of gas.

10. Utilize Avogadro's Law (PV/RT = n, at conditions of a specific pressure and temperature) to show that equal volumes of gases at the same temperature and pressure contain equal numbers of gas molecules.

11. Apply Dalton's Law of Partial Pressures, ($P_T = P_1+P_2+P_3...$), to calculate the partial pressure of a gas in a mixture or the total pressure in a mixture of gases.

12. Use Graham's Law of Diffusion, $[(R_A{}^2/R_B{}^2) = [M.W._B / M.W._A]$, to calculate either the relative rates of diffusion of two gases or the molecular weight of one of the gases.

13. Name, define and give the approximate energies involved in the three basic intermolecular forces between molecules.

14. Draw a few molecules that can participate in intermolecular hydrogen bonding and show the specific interactions involved.

15. Define boiling point and explain the factors which influence the temperature at which a liquid boils.

16. Indicate the specific make-up and the general characteristics of an a) ionic, b) molecular, c) metallic, d) polymeric, e) network and f) amorphous solid.

17. Draw a heating curve for ice, explain the main features of the curve and explain the significance of the heat of fusion and the heat of vaporization in this figure.

18. Define the important terms and comparisons in this chapter and give specific examples where appropriate.

19. Draw the phase diagram for water, pointing out the triple point and indicating the regions where only solid, liquid or gas exists.

Make a summary sheet of the important laws and corresponding equations presented in this chapter which provide the basis for characterizing the property of a gas. Indicate the parameters which are kept constant in each relationship and the units of each.

IMPORTANT TERMS AND COMPARISONS

Gaseous, Liquid and Solid State
Melting Point and Freezing Point
Barometer and Manometer
Torr, mm Hg & Pascals
Atmosphere (unit of pressure)
Systolic and Diastolic Pressure
Sphygmomanometer
Avogadro's Law
Boyle's, Charles's and Gay-Lussac's Law
Standard Temperature and Pressure (STP)
SolidsPhases and Phase Changes
Heating Curve
Heat of Fusion and Heat of Vaporization
Buckminsterfullerene or Bucky Balls
Arterial and Venous Blood
London Dispersion Force, Hydrogen Bonds and Dipole-Dipole Interactions
Fluidity
Vapor Pressure of a Liquid
Boiling Point and Melting Point
Crystallization
Ideal Gas Law
Dalton's and Graham's Laws
Crystals, Amorphous and Network
Partial and Total Pressure
Specific Heat
Phase Diagram and Triple Point
Entropy

FOCUSED REVIEW

One of the most perplexing difficulties that many students encounter in problem solving is deciding the equation to use to obtain the answer. Since there are a number of equations (relationships) which may be used in solving gas law problems, a simple procedure is necessary to find just the right relationship for the particular problem.

The parameters that may be involved in a particular problem are the following:

Parameter	Common Units
Pressure (P)	Atmosphere, mm Hg, torr, pascal
Volume (V)	Liter, milliliter
Temperature (T) ^oC, ^oK	
Moles (n)	--------
Mass (m)	Grams
Molecular Weight (M.W.)	Atomic Mass Units

The types of problems can be subdivided into two major categories. The first is the situation in which the gas is completely characterized in one state, the **conditions are CHANGED**, and one of the characteristics of the second state is unknown. A number of situations may arise.

TABLE I

	State 1 Known Quantity		State 2 Known Quantity	State 2 Unknown Quantity	
Ideal Gas Law:		$\frac{P_1V_1}{T_1} = \frac{P_2V_2}{T_2}$			
	(P_1,V_1T_1)	1.	P_2V_2	T_2	or
		2.	P_2,T_2	V_2	or
		3.	V_2,T_2	P_2	
Boyle's Law:		$P_1V_1 = P_2V_2$, at constant T			
	(P_1,V_1)	1.	P_2	V_2	or
		2.	V_2	P_2	
Charles's Law:		$\frac{V_1}{T_1} = \frac{V_2}{T_2}$, at constant P			
	(V_1,T_1)	1.	V_2	T_2	or
		2.	T_2	V_2	
Gay-Lussac's Law:		$\frac{P_1}{T_1} = \frac{P_2}{T_2}$, at constant V			
	(P_1,T_1)	1.	P_2	T_2	or
		2.	T_2	P_2	

The second category of problems involves a gas or gases in one **PARTICULAR** state. **No changes in conditions are involved.** In this case, a specific property in this state is to be calculated, using either the ideal gas law, Dalton's Law of partial pressures, Avogadro's Law or Graham's Law of diffusion.

Recall that the ideal gas law equation may be written in two forms.

$$PV = nRT \qquad (1)$$

and since

$$n = \frac{g}{M.W.}$$

$$PV = \frac{gRT}{(M.W.)} \qquad (2)$$

Therefore, one can solve for any one of the parameters in equation (1) or (2), if the other quantities are given in the problem.

Two final points of interest.

1. R, the universal gas constant is a physical quantity, which you'll recall has both a numerical value and **units**. The value of R presented in the text is R = 0.082 L-atm/mole-K. Therefore, it is essential in using this equation with this value of R that the gas parameters conform with these units of R. That is, the parameters must have units of P (atm), V (L), n (moles) and T (K). If the parameters are not given in the proper units, convert them immediately. Be on the outlook for just this situation.

2. In problems involving two states of a gas and using any one of the four gas laws presented, it is also essential that the units characterizing state 1 be the same as in state 2. For example, if the pressure in state 1 is given as 1.2 atm, while the pressure in state 2 is given as 500 mm Hg, one of these values (usually the 500 mm Hg value) must be converted to the other before the calculation.

Knowing these preliminaries and precautions, the procedure for analyzing and solving a gas law problem is:

1. Determine if the problem involves the gas in only one state or two different states.
2. Determine (**write out**) what is known and what is to be determined (unknown) in the problem.
3. From your analysis in steps 1 and 2, decide which law and equation is appropriate to solve the problem.
4. Rearrange the equation if necessary and check that all units are appropriate.
5. Solve for the unknown quantity.

To pull some of this together, consider the following problem.

If a sample of 1.78 moles of $CHCl_3$ gas occupies 68 L at a pressure of 275 mm Hg, what is its temperature in ^oC?

Step 1. The problem involves $CHCl_3$ gas in **one state**. There are no changes to a new state. Therefore the equations as written in Table I (Study Guide) are not appropriate and either the (1) ideal gas law, (2) Dalton's Law of partial pressures, (3) Avogadro' Law or (4) Graham's Law of diffusion will be used.

Step 2.

Knowns	Unknown
n = 1.78 moles	
V = 68 L	T = ?
P = 275 mm Hg	

Step 3. From the parameters given and the unknown quantity, the ideal gas law should be used in the form.

$$PV = nRT$$

Step 4. Rearrange by dividing through each side by nR. This yields

$$T = \frac{PV}{nR}$$

Convert pressure (P = 275 mm Hg) into units of atmosphere to conform with units of R, the universal gas constant.

$$P = \frac{275 \text{ mm Hg}}{1} \quad \frac{1 \text{ atm}}{760 \text{ mm Hg}} = 0.362 \text{ atm}$$

Step 5. $$T = \frac{(0.362 \text{ atm})(68 \text{ L})}{(1.78 \text{ moles})(0.082 \text{ L-atm/mole-K})} = 169 \text{ K}$$

The final step in this problem is to convert the temperature from K to ^{o}C.
$^{o}C = 273^{\circ} + K = 273^{o} + 169$,

Therefore, the final answer is: **T = 442°C**

SELF-TEST QUESTIONS

MULTIPLE CHOICE. Select the correct answer from the choices given. Only one of the choices is correct.

1. A closed aerosol-can that had been filled with shaving cream contains 3.0 g of nitrogen gas (N_2) as the propellant. If the volume of the can as 200 mL, the pressure in the can at 25°C is:

   a. 1.1 atm
   b. 130 atm
   c. 1.3 atm
   d. 13 atm

2. If the can in question 1 is thrown into an incinerator or fire, the temperature in the can could well get as high as 850 K. The pressure in the can, assuming that it did not explode, would be:

   a. 37 atm
   b. 3.7 atm
   c. 370 atm
   d. 370 mm Hg

3. A gas, which is initially contained in a movable cylinder of 900 mL volume and at 25°C., exerts a pressure of 3 atmospheres. The pressure is reduced at constant temperature to a value of 380 mm Hg. The new volume is:

a. 5400 L
b. 5.4 L
c. 7.1×10^{-3} L
d. 7.1 L

4. A piece of Dry Ice (solid carbon dioxide, CO_2) is placed in a 1 L tank. It sublimes to CO_2 gas at STP. The amount of CO_2 was:

a. 1 mg
b. 44 g
c. 4.4 g
d. 2 g

5. The average kinetic energy of molecules is proportional to the

a. temperature in °C
b. velocity of the molecule
c. temperature in K
d. volume at STP

6. The official S1 unit of pressure is:

a. torr
b. pascal
c. atmosphere
d. mm Hg

7. The autoclave operates on the principle of:

a. Dalton's Law of Partial Pressure
b. Boyle's Law
c. Gay-Lussac's Law
d. Charles's Law

8. A gas at 25°C is initially contained in a movable cylinder of 2000 mL volume and exerts a pressure of 38 mm Hg. The pressure is changed to 0.5 atm and the volume of the cylinder increases to 2.2 L. The final temperature in the cylinder is:

a. 3×10^3 K
b. 275 K
c. 3.2 K
d. 2×10^4 K

9. Consider the reaction: $3K + Y \rightarrow 2Z$. At STP, 2 L of Z are produced from the reaction of X and Y. The number of liters of X necessary is:

a. 1 L
b. 2 L
c. 3 L
d. cannot determine

10. A quantity of 6 grams of a gas occupies 300 milliliters at STP. The molecular weight of the gas is:

a. 440
b. 660
c. 44
d. 66

11. Predict which would exhibit the greatest surface tension.

a. CCl_3H
b. H_2O
c. CH_3OH
d. CCl_4

12. Indicate which of the following are characteristics of "buckyballs". There may be more than one answer.
    a. Contains 60 carbon atoms
    b. Carbons are arranged in pentagons and octagons
    c. Contains only carbon atoms
    d. Each carbon is bonded to four other carbons

13. Indicate which of the following are isoforms of carbon. There may be more than one answer.
    a. diamond
    b. soot
    c. graphite
    d. fullerenes

COMPLETION. Write the word, phrase or number in the blank which will complete the statement or answer the question.

1. STP indicates conditions in which T = __________ and P = __________.

2. The boiling point of HF (M.W. = 18) is higher than that for CH_4 (M.W. = 18) because of __________, while the boiling point of n-octane is higher than that for propane because of the __________.

3. The behavior of an ideal gas at constant volume but at varying pressure or temperature is governed by __________ Law.

4. The O_2 gas in a 1 L tank at 1 atmosphere is transferred to a 2 L tank which already contains 2 atmospheres of N_2 gas. The final pressure is __________.

5. The higher the temperature, the __________ the __________ energy of molecules. Therefore, by heating a closed container of gas, the pressure will __________.

6. A 5.6 L container of Ne or O_2 contains __________ moles of gas or __________ molecules at STP.

7. The energy required to break a hydrogen bond is approximately __________ kcal/mole.

8. A molecule used in perfume is determined to diffuse one-tenth as fast as neon gas. Its molecule weight is __________.

9. The basic postulates in the kinetic molecular theory of gases are:

    i. ______________________________

    ii. ______________________________

    iii. ______________________________

    iv. ______________________________

    v. ______________________________

    vi. ______________________________

10. The melting point of diamond, which is a __________, is much higher than that for a __________ crystal.

11. The two plateaus in a heating curve represent the temperatures at which two phases __________. The amount of heat necessary to convert one gram of water at 100°C to steam is called the __________.

12. The intermolecular forces responsible for the solidification of Xe and other noble gases are __________.

13. The unit of pressure, __________, is equivalent to the __________.

14. Convert the physical quantities in the left column to their equivalent in the right-hand column.

a.	4.3 mL	__________ L
b.	34°C	__________ K
c.	450 mm Hg	__________ atm
d.	43 g O_2	__________ moles O_2
e.	372 K	__________ °C
f.	STP	__________ mm Hg
		__________ °C
g.	17 cal/mole	__________ Kcal/mole

15. Match up the unit or instrument in the column at the right with its use in the left-hand column.

Unit or Instrument	Use
a. Barometer	i. Measure blood pressure
b. Sphygmomanometer	ii. Laboratory instrument to measure pressure
c. Closed container of liquid	iii. Hospital unit used for patients with CO poisoning, smoke inhalation, in addition to other uses
d. Hyperbaric Chamber	iv. Measures atmospheric pressure
e. Manometer	v. Used in determination of vapor pressure

16. Water has a superstructure involving intermolecular __________, which in part explains why the density of __________ is greater than that of __________.

17. As the temperature decreases, the vapor pressure of a liquid __________.

18. __________ is the transition from the solid state directly into the __________ state without going through the __________ state.

19. An amorphous solid differs from a crystalline solid in that the molecules or ions are solidified in a __________ pattern.

20. Write the equation for the behavior of an ideal gas. ______________.

21. A new form of carbon, which contains ___ carbons and has a structure which resembles a soccer ball is called either ________________ or ________________.

22. A measure of order is called _________. As the temperature of a substance increases in temperature, its _________ value will _________.

23. Examples of solids that can be described as molecular, polymeric or network are ___________, ___________, and ___________, respectively.

24. Two types of solids that exhibit very high melting points are ___________ and _________. Sodium chloride (NaCl) is an example of a __________ solid.

25. A phase diagram plots _______ on the x axis and _______ on the y axis.

26. The line in a phase diagram that separates the solid phase from the liquid phase contains all the ___________ points for that substance.

27. The _______ __________ is a unique condition in which all three phases coexist.

ANSWERS TO SELF-TEST QUESTIONS

Multiple Choice

1. d
2. a
3. b
4. d
5. c
6. b
7. c
8. a
9. c
10. a
11. b
12. a, c
13. a, b, c, d

Completion

1. T = 0°C, P = 1 atm
2. intermolecular hydrogen bonds, (higher) molecular weight
3. Gay-Lussac's Law
4. 2.5 atm
5. greater, kinetic, increase
6. 0.25, 1.5×10^{23}
7. 5-10
8. 2000
9. These statements can be paraphrased.
   i. A gas consists of molecules traveling in random directions, in straight lines and at a range of speeds.
   ii. The average kinetic energy, and therefore the average velocity of gas molecules, is directly proportional to the absolute temperature.
   iii. Gas molecules collide with each other and may exchange energy between each other, but the total kinetic energy is conserved.
   iv. The molecules in a gas take up essentially no volume. The volume of the gas is the volume of the container it is in.
   v. There are no attractions between the gas molecules.
   vi. The pressure of the gas is due to collisions of the molecules with the walls of the container. The greater the number of collisions per unit time, the greater the pressure.

10. network solid, molecular
11. coexist, heat of vaporization
12. London dispersion forces
13. torr, mm Hg or also 1 atm = 101,325 pascals is acceptable. The former equivalence is more commonly used, however.
14. a. 4.3×10^{-3} L
    b. 307 K
    c. 0.592 atm
    d. 1.3 moles O_2
    e. -1°C
    f. 760 torr, 0°C
    g. 1.7×10^{-2} kcal/mole
15. a. iv
    b. i
    c. v
    d. iii
    e. ii
16. hydrogen bonding, liquid water, ice
17. decreases
18. sublimation, gaseous, liquid
19. random
20. PV = nRT
21. 60; buckminsterfullerene; bucky balls
22. entropy, entropy, increase
23. sugar or ice; rubber, plastic or proteins; diamond or quartz
24. ionic, network, ionic
25. T, P
26. melting
27. triple point

The fluids of the human body and the seas of the earth contain common table salt, sodium chloride, and an assortment of other chemicals.

- S. Brooks

The Seas Inside Us: Water in the Life Processes

6

Solutions and Colloids

CHAPTER OBJECTIVES

After studying the chapter and working the assigned exercises in the text and study guide, you should be able to do the following.

1. Summarize the characteristics of a mixture, a suspension, a colloidal dispersion and a true solution and give specific examples of each.

2. Indicate the factors which influence the solubility of a (1) gas, (2) liquid or (3) solid in a liquid solvent.

3. Perform calculations to determine the concentration of a solution expressed in (1) molarity, (2) % w/v and (3) w/w. Given the concentration of a solution in any of these forms, convert to its equivalent in the other two.

4. Using the formula, moles = M_1 x V_1 = M_2 x V_2, calculate the concentration of solutions made by the dilution of a more concentrated solution.

5. Outline the characteristics of water which are most responsible for the unusual properties, such as (1) its high boiling point and (2) its ability to dissolve ionic compounds.

6. Specify the fundamental difference between a strong and weak electrolyte and a non-electroyte.

7. Discuss the mechanisms by which water can effectively dissolve particular covalent molecules.

8. Considering both molecular and ionic compounds, carry out calculations to determine the freezing point depression produced by the various aqueous solutions.

9. Define osmosis and osmolarity and discuss how these concepts relate to specific areas of medicine.

10. Define the important terms and comparisons in this chapter and give specific examples where appropriate.

IMPORTANT TERMS AND COMPARISONS

Homogeneous and Heterogeneous Mixtures
Solute, Solvent and Solution
Clear vs. Colorless
A Physical Constant
Strong and Weak Electrolyte
Saturated, Unsaturated and Supersaturated Solutions
"Like Dissolves Like"
Thermal Pollution
Physiological Saline
Percent Concentration
Hemolysis and Crenation
(% w/v), (w/w) and (v/v)
Osmotic Pressure
Colloidal Dispersions and Tyndall Effect
Parts per Million (ppm) and Part per Billion (ppb)
Proof
Miscible and Immisible Liquids
Plaster of Paris
Electrolyte and Non-Electrolyte
Cathode and Anode
Concentration
Solution and Suspension
Transparent, Translucent and Opaque
Solvation Layer
Dilute vs. Concentrated
Isotonic, Hypertonic and Hypotonic
Dialysis and Hemodialysis
Reverse Dialysis
Molarity and Osmolarity
Melting Point Depression
Henry's Law

FOCUSED REVIEW OF CONCEPTS

In all chemically related fields, we are continuously called upon to make observations or measurements and to characterize a mixture or solution, either in a qualitative or quantitative sense. Recall that chemistry is an experimental science and our current knowledge is derived from **careful observations** of experimental happenings. Concepts in this chapter lay the foundation for such characterization. Consider, for example, that you are presented with a test tube containing some liquid and asked to characterize it as completely as you can. Your preliminary, yet very important, judgments will necessarily be **qualitative**. Initially, you might focus on these questions.

1. Does the vial contain a true solution, a colloidal dispersion or a suspension?

2. If the vial contains a true solution,

   a. Is the solution clear (transparent), translucent or opaque?
   b. Is the solution colored or colorless?
   c. Does the solution conduct electricity?
   d. Is the liquid simply a solvent or an actual solution?
   e. If it is a solution, is it dilute or concentrated and is it unsaturated, saturated or supersaturated with solute?

At this point, it should be evident that although the characteristics are qualitative, we know a great deal about the unknown solution in the test tube.

To characterize the solution further, however, requires **quantitation**. The following table collects a number of means of quantitating solutions, with the definition and examples of each. However, first just focus your attention on a basic point that is often times not clearly defined and leads to difficulty. You will note that all the quantitative concentration terms in Table I are expressed as the quantity of one reagent in a certain volume of **SOLUTION** (= solute plus solvent). The emphasis here is to distinguish the difference between the volume of the total solution from the volume of the solvent used to dissolve the solute.

TABLE I

Quantitative Description of Solutions

Concentration Term	Definition	Examples
Molarity (M)	Moles solute per 1 L **SOLUTION** M = (moles/L)	3 M NaCl; Dissolve 175 g NaCl in enough water to make 1 L solution.
Osmolarity	Molarity times the number of particles (n) produced by each mole of solute	0.1 M NaCl; Osmolarity = 0.2 $NaCl \rightarrow Na^+ + Cl^-$; n = 2 0.3 M $Mg_3(PO_4)_2$; Osmolarity = 1.5
	Osmolarity = n x M (n = 5)	0.1 M $Mg_3(PO_4)_2 \rightarrow 3Mg^{+2} + 2PO_4^{3-}$ Osmolarity = 0.5 0.8 M glucose (n = 1) Osmolarity = 0.8
(% w/v)	Weight solute per 100 mL **SOLUTION**	5 g of X in 100 mL solution 5 % w/v
(% w/w)	Weight solute per weight **SOLUTION**	5 g of X in 50 g solution 10 % w/w
(% v/v)	In a 2 component solution, volume of A added to enough liquid B to make 100 mL **SOLUTION**	17 mL liquid A plus enough liquid B to make 100 mL solution 17 % v/v

SELF-TEST QUESTIONS

MULTIPLE CHOICE. In the following exercises, select the correct answer from the choices listed. In some cases, two or more choices will be correct.

1. Which of the following gases can dissolve in rain droplets to produce acid rain.

   a. O_2
   b. SO_3
   c. CO_2
   d. CO

2. The solubility of a solid solute in water will depend on what factors?

   a. temperature
   b. size of the solute particles
   c. polarity of solute
   d. pressure

3. Which of the following covalent molecules can hydrogen-bond to water?

a. nitrogen, N_2
b. ammonia, NH_3
c. methane, CH_4
d. ethyl alcohol, C_2H_5OH

4. Indicate which of the following properties a colloidal dispersion has that a true solution does not.

a. homogeneity
b. Tyndall effect
c. filterable with ordinary paper
d. electrical conductance

5. An example of a colloid in which a solid is dispersed in a liquid is:

a. cheese
b. whipped cream
c. jellies
d. clouds

6. Which of the following properties does a suspension **not** have?

a. settles on standing
b. homogeneous
c. translucent or opaque
d. particles size of < 1000 nm

7. A solution containing proteins and NaCl salt is placed in a semipermeable membrane bag and dialyzed against distilled water. Which component or components will be found in the distilled water?

a. NaCl, proteins
b. NaCl
c. proteins
d. none of the components

8. The solubility of a polar gas molecule in a liquid solvent increases with:

a. increasing temperature
b. decreasing temperature
c. increased gas pressure
d. increased solvent polarity

9. The preparation of 300 mL of a 2.0×10^{-5} M solution of hemoglobin (molecular weight 64,000) requires how many grams of hemoglobin?

a. 0.38 g
b. 38 g
c. 6.4×10^{-1} g
d. 7.6 mg

10. The reaction of oxygen gas with hemoglobin (Hb) is shown in the equation below:

$$Hb + 4O_2 = Hb(O_2)_4$$

How many moles of O_2 can react with the hemoglobin solution in problem 9?

a. 3×10^{-2} moles
b. 2.4×10^{-5} moles
c. 6×10^{-6} moles
d. 3×10^{-6} moles

11. A solution of HCl is prepared to be 1×10^{-2} M. A 15 mL aliquot is taken and diluted with 80 mL of water. The final concentration of HCl is

a. 1.9×10^{-4} M
b. 1.9×10^{-3} M
c. 1.6×10^{-3} M
d. 1.6×10^{-4} M

12. A 1.5 M solution is made by dissolving 50 g of compound X in 50 mL of solution. The molecular weight of X is :

a. 660
b. 100
c. 150
d. 300

13. To make a dye solution for coloring an old white shirt, a student dissolves 5 mg of dye to 10 L of water. Calculate the concentration of this solution in ppb. (Recall that 1 mL of water weighs1 g).

a. 5000 ppb
b. 50 ppb
c. 500 ppb
d. 5 ppb

14. The relative solubility of gases in water is:

a. $O_2 > N_2 > CO_2 > SO_2$
b. $SO_2 > CO_2 > O_2 > N_2$
c. $CO_2 > O_2 > N_2 > SO_2$
d. $O_2 > SO_2 > CO_2 > N_2$

15. Two liquids that are miscible are:

a. Gasoline & water
b. Oil and vinegar
c. water and alcohol
d. oil paint & water

16. Indicate which of the following is not a homogeneous solution.

a. Granite
b. CO_2 and H_2O
c. alcohol and water
d. Brass (Cu/Zn)

17. A student has 300 mL of a 2 M solution of HCl. He wants to use all of this solution and made a 0.5 M solution of HCl. Calculate how much water must he add to accomplish this.

a. 900 mL
b. 500 mL
c. 1.2 L
d. 400 mL

<u>COMPLETION</u>. Write the word, phrase or number in the blank space or draw the appropriate structure in answering the question.

1. Preparation of 500 mL of a 0.5 M NaCl solution requires __________ NaCl.

2. The osmolarity of a 0.1 <u>M</u> $Na_3(PO_4)$ solution is __________. The freezing point of this solution is determined on a 50 mL portion (aliquot) of the solution. The observed freezing point will be __________.

3. Crystalline substances that contain no water are called __________. Substances which

become hydrated on exposure to the open air are referred to as __________ compounds.

4. Three 0.1 M solutions are prepared, one with HF, another with NaCl and the last one with sucrose (common sugar). Because the labels came off the bottles, we shall identify the solutions as simply solutions A, B and C. Solutions marked A and B conduct electricity; however, solution A clearly has a much greater capacity to conduct electricity than does solution B. Solution C does not conduct electricity. Identify each solution and characterize it according to its electrolytic character.

Compound	Solution	Type of Electrolyte
HF	(i)	(iv)
NaCl	(ii)	(v)
Sucrose	(iii)	(vi)

5. A true solution contains particles with a maximum diameter of __________.

6. A __________ solution has a greater osmolarity than do red blood cells. Therefore, placing red blood cells in this solution will cause the cells to __________. A 0.3 M glucose solution is a __________(tonic) solution.

7. The concentration of a 0.40 M glucose, $C_6H_{12}O_6$, solution, can be expressed in terms of (% w/v). The equivalent concentration expressed in this way would be ______________ (% w/v).

8. The random motion of a colloid suspended in a solvent is called __________________.

9. A biochemist prepares a solution containing hemoglobin, the compound (a protein) in the red blood cells that carries O_2 to our cells. The hemoglobin in this solution is referred to as the ______________.

10. A student adds benzene to a bottle containing CCl_4 and also to a separate bottle containing methyl alcohol, CH_3OH. The benzene is miscible in the __________, but immiscible in the __________. This is an example of the rule that "like dissolves like." In this case, a ___________ compound dissolves in a ___________ solvent.

11. Osmosis involves the passage of __________ from the ____________ to the __________ side of a semipermeable membrane.

12. ____________ is the process that is used in hospitals to remove waste products from the blood of patients who have kidneys that are not functioning properly.

ANSWERS TO SELF-TEST QUESTIONS

Multiple Choice

1.	b, c	9.	a
2.	a, c	10.	b
3.	b, d	11.	c
4.	b	12.	a
5.	c	13.	c
6.	b, d	14.	b
7.	b	15.	c
8.	b, c, d	16.	a
		17.	a

Completion

1. 14.6 g
2. 0.4, -0.76°C
3. anhydrous; hygroscopic
4. (i) Solution B; (ii) Solution A; (iii) Solution C; (iv) weak electrolyte; (v) strong electrolyte; (vi) non-electrolyte
5. 1 nm
6. hypertonic, shrivel; isotonic
7. 7.2 (% w/v)
8. Brownian motion
9. solute
10. CCl_4, CH_3OH, non-polar, non-polar
11. solvent (molecules), dilute; (more) concentrated
12. Hemodialysis

Hofstadter's Law:
It always takes longer than you expect, even
when you take into account Hofstadter's Law.
- Douglas Hofstadter

7

Reaction Rates and Chemical Equilibrium

CHAPTER OBJECTIVES

After you have studied the chapter and worked the assigned exercises in the text and the study guide, you should be able to do the following.

1. Given the concentration of the reactants and/or products at various times during the reaction, calculate the rate of reaction.

2. Distinguish between molecular collisions and effective molecular collisions and how these relate the rate of a reaction.

3. Clarify how effective collisions, temperature and activation energy relate to the rate of reaction.

4. Draw and completely label the energy diagrams for (i) an endothermic reaction and (ii) an exothermic reaction.

5. Describe how a catalyst influences a reaction and its affect on the energy diagram.

6. Define the composition in the transition state (TS) and the position of the TS in the energy diagram.

7. Discuss how each of the following factors affects the rate of reaction: (1) nature of reactants, (2) concentration of reactants and products, (3) temperature and (4) the presence of a catalyst.

8. Describe the concept of dynamic equilibrium as it relates to a reversible reaction.

9. Given the equilibrium expression, write out the chemical reaction.

10. Given the chemical reaction, write out the equilibrium expression.

11. Given the concentration of the reactants and products at equilibrium, calculate the equilibrium constant, K.

12. Use LeChatelier's Principle to explain how the following factors influence the equilibrium: (1) addition of a reaction component; (2) removal of a reaction component; (3) addition of a catalyst.

13. Utilizing LeChatelier's Principle and considering both an exothermic and endothermic reaction, explain how a change in temperature (increase or decrease) will alter the equilibrium constant.

14. Define the important terms and comparisons in this chapter and give specific examples where appropriate.

IMPORTANT TERMS AND COMPARISONS

Chemical Kinetics
Initial Reaction Rate
Effective and Ineffective Collision
Activation Energy and Energy of Reaction
Transition State and Activated Complex
Energy Diagram
Endothermic and Exothermic Reactions
Endergonic and Exergonic Reactions
Rate Constant and Equilibrium Constant
Catalyst and Enzyme
Reversible and Irreversible Reaction
Dynamic Equilibrium
Haber Process
Fixation of Nitrogen
Equilibrium Constant and Equilibrium Expression
LeChatelier's Principle

SELF-TEST QUESTIONS

MULTIPLE CHOICE. In the following exercises, select the correct answer from the choices listed. In some cases, more than one choice will be correct.

1. Consider the reaction, $A + B \rightarrow C$. It was found that at 18 minutes after the start of the reaction, the concentration of C was 0.04 M. The rate of this reaction is:

   a. 450 M/min
   b. 4.2×10^{-3} M/min
   c. 900 mole/L-min
   d. 2.1×10^{-3} mole/L-min

2. Effective molecular collisions in the reaction, $A + B \rightarrow C$, are necessary for a reaction to take place because:

   a. Reactant molecules, A and B, must directly interact with each other to produce product.
   b. The collisions of rapidly moving molecules provide energy.
   c. Both a & b.

3. The activated complex in the reaction $A + B \rightarrow C + D$ is composed of:

   a. A and B
   b. C and D
   c. A
   d. A and D

4. Reaction $X + Y \rightarrow Z$ is completed in 2 hours at 20°C. The reaction is carried out at a different temperature at which the reaction is complete in 7.5 minutes. This new temperature is:

   a. 293 K
   b. 40°C
   c. 333 K
   d. 30°C

5. The equilibrium constant is equal to 5×10^{-4} for the reaction, A + B + C = 2D. The equilibrium expression for the reaction is:

a. $K = \frac{(A)\ (B)\ (C)}{(D)^2}$

b. $K = \frac{(2D)}{(A)\ (B)\ (C)}$

c. $K = \frac{(D)^2}{(A)\ (B)\ (C)}$

d. $K = \frac{(2D)^2}{(A)\ (B)\ (C)}$

6. If the equilibrium concentrations for the reactants and products in the reaction in question 5 are [A] = [B] = [C] = 1×10^{-3} M, the concentration of D would be:

a. 5×10^{-10} M
b. 5×10^{-13} M
c. 7×10^{-7} M
d. 5×10^{-5} M

7. Consider the reaction, $Cl_{2(g)} + H_2O_{(l)} = H^+ + Cl^- + HOCl$. The molarity values for of the chemical species at equilibrium are determined to be: $[Cl_{2(g)}] = 0.10$; $[H_2O] = 0.24$; $[H^+] = 1 \times 10^{-3} = [Cl^-]$; $[HOCl] = 0.4$. The equilibrium constant for this reaction is:

a. 1.7×10^{-5}
b. 1.7×10^{-2}
c. 6×10^{4}
d. 6×10^{2}

8. Which of the following describe the Haber process?

a. $N_{2(g)} + 3H_{2(g)} = 2NH_{3(g)}$
b. Exothermic
c. Uses a catalyst
d. All of the above

9. If one monitors the rate of the formation of C in the reaction, $A + B \rightarrow C$, one finds the rate is 5×10^{-2} M/min. After one hour of reaction, the concentration of C is:

a. 5×10^{-2} M
b. 3 M
c. 8×10^{-4} M
d. 30 M

10. Indicate the statement that correctly describes the relationship between the equilibrium constant, K, and the rate of the reaction.

a. The larger the K value, the faster the rate of the reaction is.
b. The larger the K value, the smaller the rate of reaction is.
c. There is no simple relationship between K and the rate of reaction.
d. The faster the initial rate of the reaction, the larger the K value.

COMPLETION. Write the word, phrase or number in the blank space that completes the statement.

1. Indicate whether the following reactions are endothermic or exothermic.

a. $A + B + \text{Energy} \rightarrow C$ ______________________________

b. $A + B \rightarrow C + D + \text{Energy}$ ______________________________

c. $A + B \rightarrow C$
$\Delta E = -40$ kcal/mole ______________________

d. $H_{2(g)} + O_{2(g)} \rightarrow H_2O_{(e)}$ + Energy ______________________

2. Indicate how each of the following will affect the rate of formation of C in the reaction, $A + B = C$. The answer should be increase, decrease or no effect (on rate).

a. Add a catalyst ______________________
b. Increase concentration of A ______________________
c. Decrease concentration of B ______________________
d. Decrease the temperature ______________________

3. Consider the exothermic reaction, $A + B \rightarrow C + D$, -10 kcal, in which the activation energy for the reaction is 25 kcal. <u>Complete</u> the diagram below with labels and showing all items of importance.

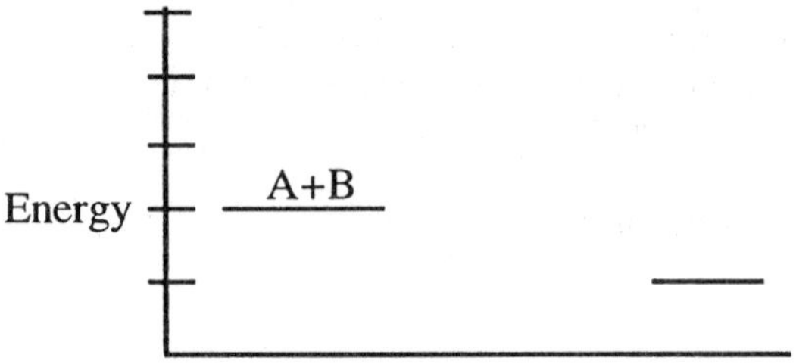

4. Completely label the following diagram by placing the appropriate label in each indicated block.

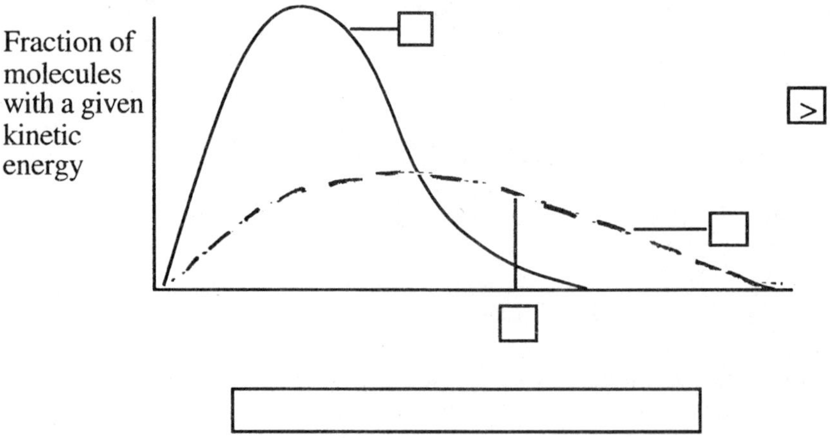

5. Molecules that act as catalysts in the human body are called ______________.

6. Catalysts in a reaction lower the ______________ of the reaction by increasing the number of ______________ between reactant molecules, without increasing the reaction temperature. However, the energy of the reactants and products (do or do not)______ change by the presence of a catalyst.

7. Complete the following table.

Equilibrium Expression	Reaction
a.	$H_2 + O_2 = H_2O$
b. $\frac{(W)^2}{(A)(B)} = K$	
c.	$H_2O = H^+ + OH^-$

8. The concept of a dynamic equilibrium brings out the finding that although the _______________ of both reactant and product molecules (does or does not) change, both the _______________ and _______________ reactions are taking place.

9. Using LeChatelier's Principle, indicate how the following changes will affect the equilibrium in the exothermic reaction. Indicate whether the reaction goes to the right or to the left or has no change.

$$A + B = C$$

Change	Effect
a. Increase [A]	
b. Decrease [C]	
c. Add a catalyst	
d. Increase temperature	

ANSWERS TO SELF-TEST QUESTIONS

Multiple Choice

1. d
2. c
3. a and b; note that A and B are the only reactants and C and D are the only products. The atoms in A and B are the same ones in C and D, only rearranged to form the new molecules.
4. c
5. c
6. c
7. a
8. d
9. b
10. c

Completion

1. a. endothermic
   b. exothermic
   c. exothermic
   d. exothermic
2. a. increase rate
   b. increase rate
   c. decrease rate
   d. decrease rate (number of effective collisions is reduced

3.

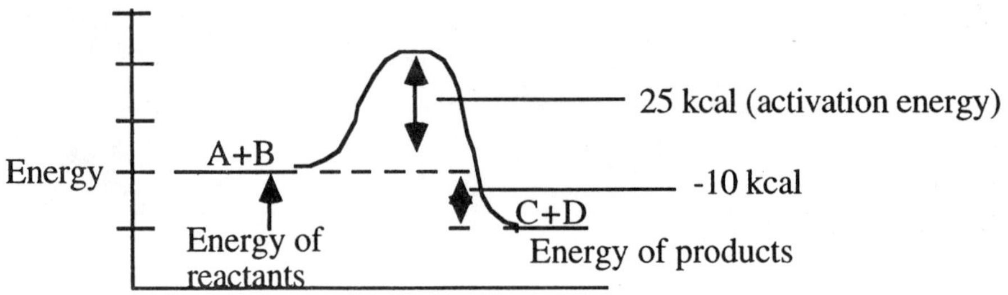

4.

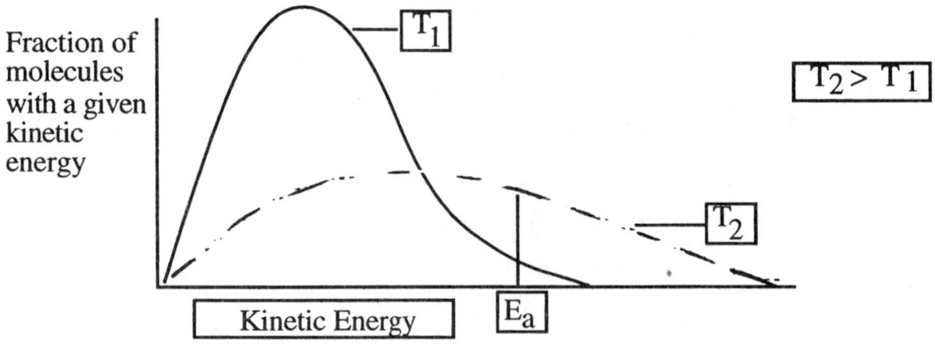

5. enzymes
6. activation energy, effective collisions, do not
7. a. $K = \dfrac{(H_2O)}{(H_2)(O_2)}$

   b. A + B = 2 W

   c. $K = \dfrac{(H^+)\,(OH^-)}{(H_2O)}$

8. concentrations, does not, forward, reverse
9. a. reaction goes to right
   b. reaction goes to right
   c. no effect
   d. reaction goes to left. In an exothermic reaction, heat can be considered a product. Therefore, increasing the temperature will increase the heat (a product) and the reaction will go to the left to relieve the stress.

Concerning solutions to acid-base equilibrium problems: The answers to the riddle of life are not found in the back of the book.
- Charlie Brown

8

Acids and Bases

CHAPTER OBJECTIVES

As will become evident in this chapter and those focused on biochemistry (Chapters 19-31), the body contains aqueous systems vital to life as we know it. These systems include intracellular fluids, blood, gastric juices, saliva and more. The importance of understanding the **CHARACTERISTICS AND REACTIONS** of **ACIDS, BASES** and **SALTS** in aqueous solutions therefore cannot be overemphasized. This chapter dwells on these fundamentals.

After you have studied the chapter and worked the assigned exercises in the text and the study guide, you should be able to do the following.

1. Characterize Arrhenius acids and bases and define the similarities and differences with Bronsted-Lowry acids and bases. Give specific examples of each.

2. Explain the difference between strong and weak acids (bases) and give four examples of strong acids (bases).

3. Write the reactions involving Bronsted-Lowry acids and bases and point out the conjugate acid-base pairs.

4. Define Ka and pKa values and explain how they are related to acid strength.

5. Write ionic equations for the neutralization reaction of acids and bases; clearly define the salt that forms in each case.

6. For an aqueous solution, define the neutral point and the acidic and basic domains in terms of pH, pOH, $[H^+]$ and $[OH^-]$. At any pH, show the relationship between pH and pOH and also $[H^+]$ and $[OH^-]$.

7. Given the $[H^+]$ or $[OH^-]$ of a solution, calculate the pH. Given the pH or pOH of a solution, calculate the $[H^+]$.

8. Name six solutions that are body fluids, drinks, or common household products which are (1) acidic or which are (ii) basic. In each case, indicate its approximate pH value.

9. Explain the importance of acid-base indicators in qualitatively testing the acidic or basic character of a solution.

10. Make a list of salts and characterize them as either neutral, acidic or basic salts.

11. Write equations for the hydrolysis reactions associated with acidic and basic salts. Indicate whether the resulting solution is acidic or basic.

12. Define buffer solution and indicate the components and their relative concentrations necessary to prepare buffers with good buffering capacity.

13. Explain the importance of the Ka and pKa value in determining the most appropriate acid to use to prepare a buffer solution.

14. Show how the components of carbonate and phosphate buffers react on addition of strong acid or base.

15. Explain in some detail the way in which you would determine the concentration of a known volume of basic solution by titration with a strong acid of known concentration.

16. Name a few acids that have more than one titratable proton. Explain how this must be taken into account in calculations for an acid-base titration.

17. Write out the relationship in the Henderson-Hasselbalch equation and indicate types of problems in which this is used.

18. Define the important terms and comparisons in this chapter and give specific examples where appropriate.

IMPORTANT TERMS AND COMPARISONS

Arrhenius Acid and Base
Hydronium Ion and Hydroxide Ion
Strong and Weak Acid
Mon-, Di- and Triprotic Acids
Glacial Acetic Acid
Bronsted-Lowry Acid and Base
Basic, Acidic or Neutral Salts
Acid Ionization Constant, Ka
Acid-Base Titration
pKa, pH and pOH
Buffer Solutions
Hyper- and Hypoventilation
Metal Oxide and Hydroxide
Baking Soda
Acidosis and Alkalosis

Ion Product of Water, Kw
Acidic and Basic Solution
Amphiprotic Substance
pH Indicator
pH Paper and pH Meter
Conjugate Acid-Base Pair
Hydrolysis
Mono-, Di- and Triprotic Acid
Buret
Ionic Equation
Buffering Capacity
Antacid
Pipet
Henderson-Hasselbalch Equation

SELF-TEST QUESTIONS

<u>MULTIPLE CHOICE.</u> In the following exercises, select the correct answer from the choices listed. In some cases, more than one choice will be correct.

1. The pH of a 1 x 10^{-4} M NaOH solution is

    a. 4 c. 10
    b. 1 d. 6

2. Which of the following is (are) strong acid(s)?

    a. perchloric acid c. hydroiodic acid
    b. boric acid d. acetic acid

3. Which of the following is both a strong acid and an oxidizing agent?

    a. NH_3 c. HNO_3
    b. H_2SO_4 d. H_2O

4. Indicate which of the following characteristics is (are) inherently associated with a strong acid solution.

    a. concentrated c. dissociates close to 100%
    b. low pH value d. reacts with bases

5. The species in a phosphate buffer at pH 6.5 that reacts with added hydroxide ion is:

    a. $H_2PO_4^-$ c. HPO_4^{-2}
    b. H_3PO_4 d. PO_4^{-3}

6. A 100 mL solution of 0.1 M phosphoric acid is titrated with a 0.1 M NaOH solution. To reach the equivalence point will require the addition of what volume of NaOH?:

    a. 100 mL c. 300 mL
    b. 33.3 mL d. 200 mL

7. The Ka value for lactic acid is 1.4 x 10^{-4}. The pKa value is:

    a. 4.1 c. 4.6
    b. 3.8 d. 3.1

8. A solution with pH = 7.7 has a $[H^+]$ of:

    a. 2 x 10^{-7} M c. 2 x 10^{-8} M
    b. 7 x 10^{-7} M d. 7 x 10^{-8} M

9. The pKa values for four acids are 2.6, 3.6, 4.9 and 5.2. The weakest acid is the one with pKa value of:

    a. 2.6 c. 4.9
    b. 3.6 d. 5.2

10. An aqueous solution made up of 0.1 M sodium acetate ($NaCH_3CO_2$) will be:

a. acidic
b. basic
c. neutral
d. cannot determine

11. The addition of strong acid to a solution that is initially at pH 8 does not change the pH. The original solution:

a. contains a strong acid
b. contains an indicator
c. is a buffer solution
d. is past its equivalence point

12. Consider the reaction: $PO4^{-3} + H2S = HPO4^{-2} + HS^{-}$
Refer to Table 8.2 in the text for a listing of acids and their conjugate bases.
The position of the equilibrium;

a. lies toward the reactants
b. lies toward the products
c. cannot be predicted
d. will have n equilibrium constant of K =1

13. The pOH of a solution made up of 2×10^{-4} moles of $Mg(OH)_2$ in 400 mL of water is:

a. 11
b. 12
c. 3
d. 2

14. Considering the acids below, together with the corresponding conjugate bases, which buffer system would be used to produce a buffer solution at pH 9.0?

a. acetic acid, $K_a = 1.8 \times 10^{-5}$
b. boric acid, $K_a = 7.3 \times 10^{-10}$
c. formic acid, $K_a = 1.8 \times 10^{-4}$
d. carbonic acid, $K_a = 4.3 \times 10^{-7}$

15. The Henderson-Hasselbalch equation is used to calculate which of the following:
a. whether an acid is strong or weak
b. whether a reaction between an acid and base will go to completion
c. the extent of acidosis or alkalosis in the blood
d. the ratio of the concentrations of the acid and its conjugate base to be used in preparing a buffer solution at the desired pH.

16. A buffer may be prepared by mixing equal amounts of a

a. weak acid and its conjugate base
b. weak acid and NaOH
c. strong acid and its conjugate base
d. strong acid and a weak base

17. A solution of acetic acid (pK = 4.75) is titrated with NaOH until the ratio of [acetate]/[acetic acid] = 10. Use the Henderson Hasselbalch, calculate the pH of the solution.

a. 3.75
b. 5.75
c. 4.75
d. 5.00

COMPLETION. Write the word, phrase or number in the blank space in answering the question.

1. Complete and balance the following reactions.

   a. $HCl + NaOH \rightarrow$ ________ + ________ + ________

   b. $Na_2O + HNO_3 \rightarrow$ ________ + ________ + ________

   c. $H_2SO_4 +$ ________ $\rightarrow K_2SO_4 + 2H_2O$

   d. $NH_3 + HCl \rightarrow Cl^- +$ ________

   e. $Na_2CO_3 + HI \rightarrow$ ________ + ________ + ________
   final products

2. A 100 mL sample of unknown acid is titrated with 0.42 M NaOH. The end-point of the titration requires 29 mL of base. The molarity of the acid is ____________________.

3. If the acid in question 2 were H_2SO_4, the molarity of the H_2SO_4 would be ________________.

4. Indicate (yes or no) whether the following bases are Arrhenius and/or Bronsted-Lowry bases.

	Arrhenius	Bronsted-Lowry
a. NH_3		
b. NaOH		
c. $Al(OH)_3$		

5. The term hydronium ion is used interchangeably with ____________, ____________ and ________________.

6. The ______________ of an acid in an intrinsic property of an acid, while the concentration is a variable and depends on the preparation of the solution.

7. Indicate the conjugate acid or base to the species below.

   a. H_3O^+ ________________________ (conjugate base)

   b. H_2O ________________________ (conjugate base)

   c. HSO_4^- ________________________ (conjugate acid)

   d. Cl^- ________________________ (conjugate acid)

8. Indicate whether the acids shown below are mono-, di- or triprotic acids.

   a. H_3PO_4 ________________________

b. H_3CCOOH ______________________

c. H_2S ______________________

d. H_2CO_3 ______________________

e. $H_2PO_4^-$ ______________________

9. Complete the following table

	$[H^+]$	[OH]	pH
a.	1×10^{-2} M	______________________	______
b.	_________	2.5×10^{-3} M	______
c.	_________	______________________	6.4

10. Indicate whether the dissolution of the following salts will produce a neutral, acidic or basic solution.

a. KCl ______________________

b. NH_4I ______________________

c. $Na(HCO_2)$ (formic acid is a weak acid) ______________________

d. $NaClO_4$ ______________________

11. If the pH of the blood goes below the normal pH, the condition is called ______________. This may be caused by ____________________________, a difficulty in breathing.

ANSWERS TO SELF-TEST QUESTIONS

Multiple Choice

1. c
2. a, c
3. b, c
4. c, d
5. a
6. c
7. b
8. c
9. d
10. b
11. c
12. b
13. c
14. b
15. d
16. a
17. a

Completion

1. a. H_2O, Na^+, Cl^-
   b. H_2O, 2 Na^+, 2 NO_3^-
   c. 2 KOH
   d. NH_4^+
   e. H_2O, $CO_2\uparrow$, 2 NaI

2. 0.12 M

3. 0.06 M

4. a. no, yes
   b. yes, yes
   c. yes, yes

5. proton, H^+, H_3O^+

6. strength

7. a. H_2O
   b. OH^-
   c. H_2SO_4
   d. HCl

8. a. triprotic
   b. monoprotic
   c. diprotic
   d. diprotic
   e. diprotic

9. a. 1×10^{-12} M, 2

   b. 4.0×10^{-12}, 11.4

   c. 4×10^{-7}, 2.5×10^{-8}

10. a. neutral; salt of a strong acid (HCl) and a strong base (KOH)
    b. acidic, salt of a strong acid (HI) and a weak base (NH_4OH)
    c. basic; salt of a weak acid (HCO_2H) and a strong base (NaOH)
    d. neutral; salt of a strong acid ($HClO_4$) and a strong base (NaOH)

11. acidosis, hypoventilation

Every man, woman and child inhabiting this planet became an unwitting guinea pig in a vast biochemical experiment on July 16, 1945, when the first atomic bomb was successfully detonated.

- Ernest Borek
The Atoms Within Us

9

Nuclear Chemistry

CHAPTER OBJECTIVES

After you have studied the chapter and worked the assigned exercises in the text and the study guide, you should be able to do the following.

1. Characterize the three major types of radioactivity in terms of symbol, charge, mass and penetrating capability.

2. Associate the names of famous scientists with their accomplishment in nuclear chemistry.

3. Given the frequency (υ) of electromagnetic radiation and the equation (λ) (υ) = c, calculate the wavelength (λ). Given the wavelength, calculate the frequency.

4. Order the different forms of radiation in the electromagnetic spectrum in terms of increasing energy and frequency and decreasing wavelength.

5. Explain the difference between a chemical and a nuclear reaction.

6. Write balanced nuclear equations for a reaction involving (i) an alpha particle, (ii) a beta particle and (iii) a gamma emission. Note whether the product nucleus has a mass greater than, equal to or less than the original radioactive nuclide.

7. Given an equation that includes all but one particle in a nuclear reaction, balance the equation and identify the missing particle.

8. Explain the term "half-life" and relate this to the decay curve for a radioactive nuclide.

9. Explain how a Geiger-Muller counter functions in monitoring the intensity of beta particle emission.

10. Define the three units, roentgens, rems and rads, used to describe the effects of radiation and indicate how they relate to each other.

11. Explain how ionizing radiation damages tissue by the formation of free radicals.

12. Write out the relationship that describes how the intensity of ionizing radiation varies with distance.

13. Describe the role of cobalt-60, iodine-131, and technetium-99m in medical diagnosis and/or treatment.

14. Explain the importance of fusion reactions to life on earth.

15. Point out the most obvious differences between nuclear fusion and fission reactions. Outline how the fission reactions can be used to produce controlled nuclear energy on the one hand and the atomic bomb on the other hand.

16. Define the important terms and comparisons in this chapter and give specific examples where appropriate.

IMPORTANT TERMS AND COMPARISONS

Alpha (α) Particle, Beta (β) Particle, and Gamma (γ) Ray
Equivalent Dose
Wavelength (λ) and Frequency (υ)
Radiation (Exposure) Badge
Photon and electron
Nuclide
Background Radiation
Tagging a Drug
Mass Number and Atomic Number
Hydrogen, Deuterium and Tritium
Isotope
Alpha, Beta or Gamma Emitter
$E = mc^2$
Isotope Generator
Half-Life ($t_{1/2}$)
Curie
Nuclear Power Plant
Control Rods
Nuclear Waste
Breeder Reactor
$(I_1)(d_1{}^2) = (I_2)(d_2{}^2)$
Positron Emission Tomography (PET)
Roentgen, Rad and Rem
Transmutation
Electromagnetic Spectrum
$(\lambda)(\upsilon) = c$
Mutation
Subatomic Particles
Ground State
Hyperthyroidism
Collimated Radiation Beam
Radiation Sickness
"Big Bang" Explosion
Fusion
Ionizing radiation
Natural and Artificial Transmutation
Scintillation Counter
Radioactive Decay Curve
Nuclear Fission and Fusion
"Phosphor"
Geiger-Muller Counter
Spectrum of Radiation
Computer-Assisted Tomography (CAT)
Magnetic Resonance Imaging (MRI)

SELF-TEST QUESTIONS

MULTIPLE CHOICE. In the following exercises, select the correct answer from the choices listed. In some cases, more than one choice will be correct.

1. A 4 g sample of a radionuclide at room temperature has a half-life of three months. The amount of sample remaining after a year is:

a. 0.25 g
b. 0.40 g
c. 2 g
d. 4 g

2. If the radioactive decay of the sample in exercise 1 were carried out -100°C, the decay process would:

a. decrease
b. increase
c. remain unchanged
d. cannot determine

3. A nucleus which emits a beta particle will be transformed into a product nucleus of the:

a. same mass number
b. same atomic number
c. atomic number one unit greater
d. mass number one unit greater

4. The wavelength of blue light is 4000 A (1 A = 10^{-8} cm). Its frequency is:

a. $12 x 10^{-12}$ s^{-1}
b. $7.5 x 10^{14}$ s^{-1}
c. $7.5 x 10^{10}$ s^{-1}
d. $12 x 10^{10}$ s^{-1}

5. A scintillation counter uses a "phosphor" to:

a. determine $t_{1/2}$ values
b. produce transuranium elements
c. measure the number of mutations
d. measure the radioactivity in a sample

6. A millicurie is a unit of radioactivity equivalent to:

a. $3.7 x 10^{10}$ counts s^{-1}
b. $3.7 x 10^{10}$ counts min^{-1}
c. $3.7 x 10^{7}$ counts s^{-1}
d. $3.7 x 10^{4}$ counts s^{-1}

7. The radiation that causes the most damage within the cell is:

a. alpha particles
b. beta particles
c. gamma rays
d. X-rays

8. Which of the following radioactive nuclides decay the fastest?

a. $^{197}_{80}Hg$, $t_{1/2}$ = 65 hrs
b. $^{210}_{84}Po$, $t_{1/2}$ = 138 days
c. $^{238}_{92}U$, $t_{1/2} = 4x10^{9}$ yr.
d. $^{131}_{53}I$, $t_{1/2}$ = 8 days

9. The wavelength of which radiation is the greatest (longest)?

a. microwaves
b. X-rays
c. visible light
d. radiowaves

10. The number of neutrons in phosphorus-32, $^{32}_{15}P$ is:

a. 32
b. 15
c. 17
d. 5

11. Which of the following nuclear reactions does not involve natural transmutation?

a. $^{11}_{5}B \rightarrow {}^{11}_{5}B + \gamma$

b. $^{210}_{84}Po \rightarrow {}^{206}_{82}Pb + {}^{4}_{2}He$

c. $^{14}_{6}C \rightarrow {}^{14}_{7}N + {}^{0}_{-1}e$

12. A source of ionizing radiation is found to have an intensity of 25 mCi at a distance of 1 m. To reduce the intensity of this radiation to 1 mCi, calculate what distance would be needed.

a. 5 m
b. 25 m
c. 2 m
d. 10 m

COMPLETION. Write the word, phrase or number in the blank space in answering the question.

1. Consider the reactions in question 11 in the multiple choice section. Reaction a) is an example of a(n) ______________ emitter, reaction b) shows a(n) ______________ emitter, while reaction c) exhibits a(n) ______________ emitter.

2. Complete the following nuclear reactions. Refer to a periodic table.

a. $^{28}_{13}Al \rightarrow \beta +$ ______________

b. ______________ $\rightarrow {}^{1}_{1}H + {}^{0}_{-1}e$

c. $^{254}_{102}No \rightarrow {}^{4}_{2}He +$ ______________

d. $^{87}_{36}Kr \rightarrow {}^{87}_{37}Rb +$ ______________

3. The order of increasing penetrating power for gamma rays, alpha and beta particles is:

______________ $<$ ______________ $<$ ______________

4. Match up the names of the famous scientist in the left-hand column with accomplishments or subjects most associated with the individual.

a.	H. Becquerel	(i)	Discovered X-ray
b.	E. Rutherford	(ii)	$E = mc^2$
c.	G. Seaborg	(iii)	CAT scan
d.	A. Einstein	(iv)	Prepared transuranium elements
e.	W. Roentgen	(v)	"Coined" the names alpha, beta and gamma rays
f.	A. Cormack and G. Hounsfield	(vi)	3.7×10^{10} counts s^{-1}
g.	M. Curie	(vii)	Observed that certain rocks gave off mysterious radiation

5. Radiation that is derived from both cosmic rays and rocks in our environment contribute to the so-called ______________ radiation.

6. Another name commonly used for an alpha particle is the ______________ of a ______________ atom.

7. A beta particle has mass number of ______________ and an atomic number of ______________.

8. Cosmic rays are streams of ______________, while gamma rays consist of ______________, ______________.

9. Isotopes have the same number of ______________ but different number of ______________.

10. A sample of a radioactive nuclide with a half-life of 10 years is left in a cave for 35 years. After this time, only 0.5 grams remains. The amount of this radioactive material originally left in the cave was ______________.

11. A sample of radioactive isotope has an activity of 4.0 mCi. This sample emits ______________ counts per second.

12. Diagnostic X-rays are absorbed by the bones and not the ______________ because the bones contain elements of ______________ atomic number which absorb X-rays strongly.

13. Consider the following two reactions:

a. $^{252}_{98}Cf + ^{10}_{5}B \rightarrow ^{257}_{103}Lr + 5\ ^{1}_{0}n$

b. $^{235}_{92}U + ^{1}_{0}n \rightarrow ^{90}_{37}Rb + ^{144}_{55}Cs + 2\ ^{1}_{0}n$

Reaction (a) is an example of a ______________ reaction, while reaction (b) is a ______________ reaction.

14. An important characteristic of a breeder reaction is that it not only provides ______________, but also makes additional ______________.

15. A ______ __________ is a compound with __________ electrons and as a result is highly reactive.

16. Medical diagnosis has recently made significant advances in that physicians can diagnose many medical problems by non-invasive procedures (no surgical procedures). Complete the following table that characterizes three of these procedures.

Acronym	Complete Name	Probe (Radiation) Used In Diagnosis
CAT (scan)		
PET (scan)		
MRI		

ANSWERS TO SELF-TEST QUESTIONS

Multiple Choice

1. a
2. c
3. a, c
4. b
5. d
6. c
7. a
8. a
9. d
10. c
11. a
12. a

Completion

1. a. gamma
   b. alpha
   c. beta

2. a. ${}^{28}_{14}Si$
   b. ${}^{1}_{0}n$
   c. ${}^{250}_{100}Fm$
   d. ${}^{0}_{-1}e$

3. alpha < beta < gamma

4. a. (vii)
   b. (v)
   c. (iv)
   d. (ii)
   e. (i)
   f. (iii)
   g. (vi)

5. background

6. nucleus, helium

7. zero, minus one

8. nuclei (mainly protons), high energy radiation

9. protons, neutrons

10. ca. 6 g

11. 1.5×10^8

12. soft tissue, higher

13. fusion, fission

14. energy, fuel

15. free radical; unpaired

16.

CAT (scan)	Computer-Assisted Tomography	X-rays
PET (scan)	Positron Emission Tomography	Isotopes emitting positrons
MRI	Magnetic Resonance Imaging	Radio-frequency waves

Gasoline - 29¢/gal
- Sign at gas station in Forty Fort, PA; circa 1965

10

Organic Chemistry

CHAPTER OBJECTIVES

After you have studied the chapter and worked the assigned exercises in the text and the study guide, you should be able to do the following.

1. Indicate the importance of the experiment that F. Wohler carried out in which he produced urea from ammonium chloride and silver cyanate.

2. Compare and contrast the properties of organic and inorganic compounds.

3. Review the octet rule and how to draw Lewis structures for simple organic molecules, as outlined in Chapter 3.

4. Review the valence-shell electron pair repulsion theory (VSEPR) in Chapter 3 and predict the structures and bond angles in simple organic molecules.

5. Be able to recognize and structurally define the different organic functional groups. The major functional groups include alcohols, amines, aldehydes, ketones, and carboxylic acids.

6. Define the difference between a primary, secondary and tertiary alcohol and be able to recognize and draw each of them.

7. Distinguish between primary, secondary and tertiary amines and be able to draw each of them.

8. Recognize the difference between an aldehyde, ketone and carboxylic acid functional group, all of which contain a carbonyl group, and be able to draw each of them

9. Define the important terms and comparisons in this chapter and give specific examples where appropriate.

IMPORTANT TERMS AND COMPARISONS

Lewis Structures
Organic Chemistry
Functional Groups
Alcohols, Aldehydes, Ketones, Carboxylic Acids and Amines
Condensed Structural Formulas
AlcoholsPrimary, Secondary and Tertiary Amines
VSEPR Theory
Synthetic and Natural Compounds
Molecular and Structural Formula
Primary, Secondary and Tertiary

SELF-TEST QUESTIONS

MULTIPLE CHOICE. In the following exercises, select the correct answer from the choices listed. In some cases, more than one choice will be correct.

1. The molecular formula, C_3H_8O, may represent the following class(es) of organic compound(s). Indicate which one is correct.

   a. aldehyde
   b. carboxylic acid
   c. amine.
   d. alcohol

2. Indicate which of the following Lewis structures is correct.

   a. H-Ö-H

   b. :C≡O:

   c. 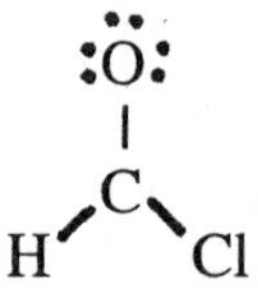

   d. 

3. Indicate what type of amine the molecule, ethylamine, is. $H_3C\text{-}CH_2\text{-}N\text{-}H$ (with H bonded below N)

   a. primary amine
   b. secondary amine
   c. tertiary amine
   d. quaternary amine

4. A tertiary carbon is bonded directly to:

   a. 2 hydrogens
   b. 3 carbons
   c. 2 carbons
   d. 4 carbons

5. The angle between the carbon atoms in propane, shown below, is

   $H_3C\text{-}CH_2\text{-}CH_3$

   a. 180^0
   b. 109.5^0
   c. 120^0
   d. 100^0

6. The bonding in methane (CH_4) involves sp^3 hybrid orbitals. Which of the following is true about the orbital and the bonding?

   a. each hybrid orbital has a 50% contribution from the s and the p orbital
   b. each hybrid orbital is a mixture of one part s orbital and three parts p orbital
   c. the angle between the hybrid orbital is 109.5 degrees.
   d. there are four sp^3 hybrid orbitals formed

7. Indicate which two elements make up about 75 % of the earth.

   a. H and C
   b. C and Fe
   c. C and Si
   d. O and Si

8. Indicate which of the following properties is associated with organic compounds. There may be more than one answer.

   a. soluble in water
   b. most of them will burn
   c. they conduct electricity
   d. bonding is almost entirely covalent

9. Indicate which of the following properties is associated with inorganic compounds. There may be more than one answer.

   a. many conduct electricity
   b. reactions are very fast
   c. C is the major component in these compounds
   d. most are solids with high melting points

<u>COMPLETION</u>. Write the word, phrase or number in the blank space or draw the appropriate structure in answering the question.

1. A simple aldehyde molecule contains a ____________ group, with the requirement that the (carbonyl) carbon atom must be the ______________ carbon atom in the molecule.

2. A simple ketone contains a _____________ group and it must be positioned at an __________ position in the molecule.

3. Cholesterol is made in the liver (natural) and also has been made synthetically in the chemical laboratory. A comparison of the structure and its chemical characteristics shows that the cholesterol made by these entirely different routes is ______________.

4. Draw the condensed structural formula for a primary and a tertiary amine that has three carbons and the necessary number of hydrogens.

5. Predict the HCH, HCC and the OCH bond angles in the ketone, acetone.

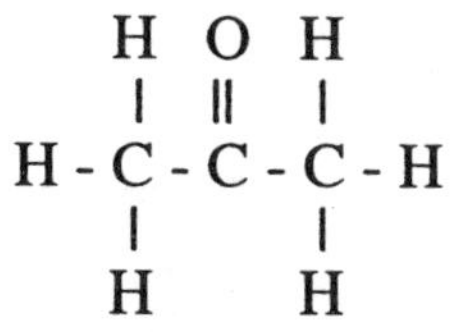

Bond angle

a. HCH ____________

b. HCC ____________

c. OCH ____________

6. Draw the Lewis structures for the following molecules.

a. HCl

b. formic acid, HCOOH

c. H_3CCH_2CHO

d. $H_2NCH_2CH_3$

7. Identify the functional groups in the following molecules.

a. 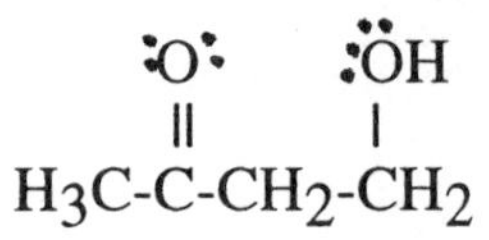

b. 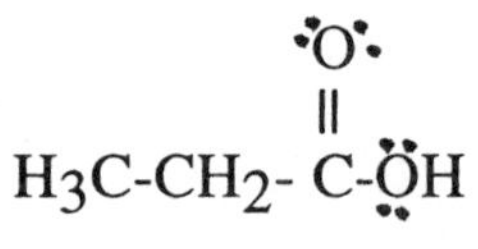

c. 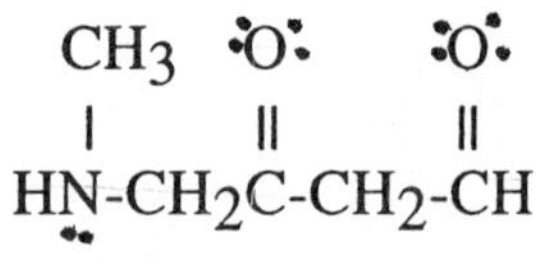

8. Draw at least four condensed structural formulas for the amines with the molecular formula of C_4NH_{11}.

9. Indicate the hybridization of the carbon atom in the compounds below and indicate the bond angle that is associated with that hybridization.

a. $R_2C=\ddot{O}$

b. CCl_4

c. $H-C\equiv C-H$

10. Draw the condensed structure for a compound with the molecular formula indicated.

a. a secondary alcohol with formula, C_3H_8O.

b. a tertiary amine with formula, $C_4H_{11}N$.

c. a molecule that has two aldehyde groups (a dialdehyde) with the formula, $C_4H_6O_2$.

ANSWERS TO SELF-TEST QUESTIONS

Multiple Choice

1. only an alcohol
2. b, d
3. a
4. b
5. b. It is important to understand that in simple drawings, the carbons appear in a line (180^0); however, the (CCC) angles are 109.5^0.
6. b, c, d
7. d
8. b, d
9. a, b, d

Completion

1. carbonyl, terminal or end
2. carbonyl, internal
3. identical
4.

primary amine $H_3CCH_2CH_2\ddot{N}H_2$

tertiary amine $H_3C\ddot{N}CH_3$
with CH_3 attached to N

5. a. 109.5^0
   b. 109.5^0
   c. 120^0

6. a. 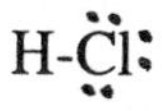

b. 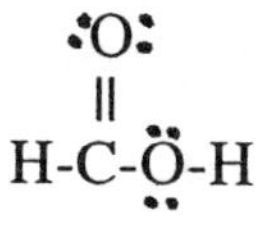

c. 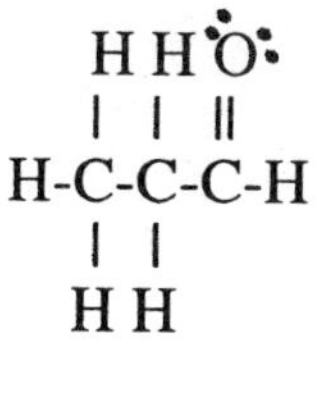

d. 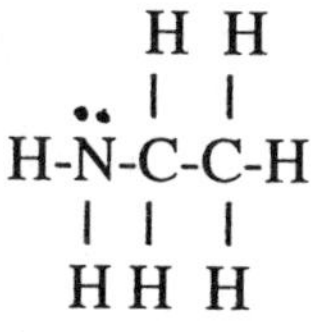

7. a. ketone and alcohol
   b. carboxylic acid
   c. ketone, aldehyde and amine

8.

H
$H_3CCH_2CH_2\underset{\cdot\cdot}{N}CH_3$

$H_3CCH_2\ddot{N}[(CH_3)]_2$

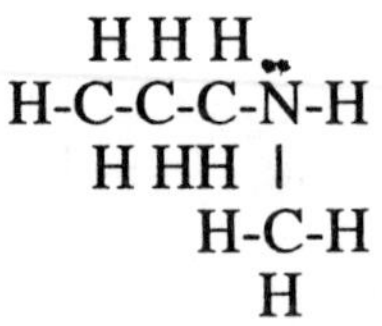

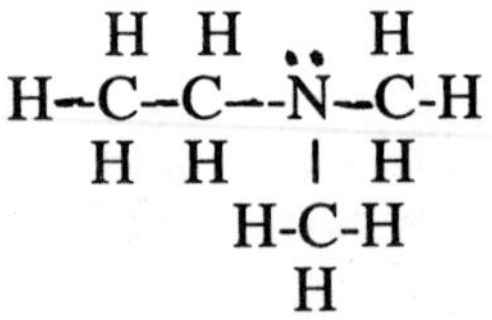

$[(CH_3)]_2\ddot{N}CH_2CH_3$

$[(CH_3)]_2CH\ddot{N}H(CH_3)$

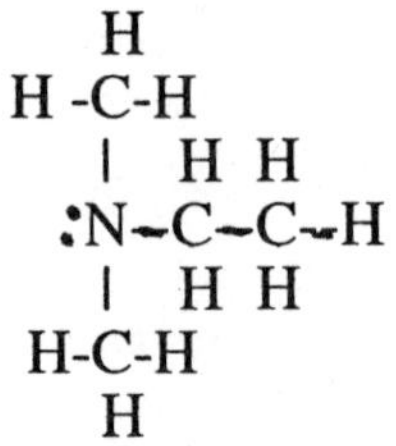

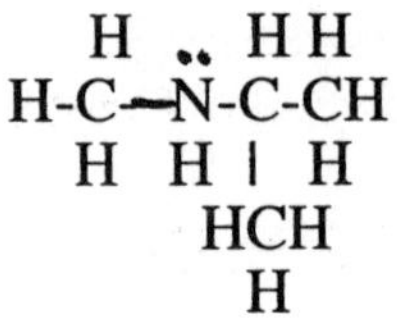

9. a. sp^2 hybridization; 120^0
   b. sp^3 hybridization; 109.5^0
   c. sp hybridization; 180^0

10.

a.
:OH
|
CH_3CHCH_3

b $(CH_3)_2\ddot{N}CH_2CH_3$

c. $CH\ddot{\underset{\cdot\cdot}{O}}CH_2CH_2CH\ddot{O}$:

Gasoline - 29¢/gal
- Sign at gas station in
Forty Fort, PA; circa 1965

11

Alkanes

CHAPTER OBJECTIVES

After you have studied the chapter and worked the assigned exercises in the text and the study guide, you should be able to do the following.

1. Name and draw the first ten unbranched alkanes, noting trends in their physical characteristics.

2. Become proficient in drawing all members of the alkane family using the abbreviated line-angle formula.

2. Consider the molecular formula for some alkanes with greater than 4 carbons and identify any constitutional or structural isomers.

3. Using the IUPAC rules of nomenclature for saturated hydrocarbons, write the names for a variety of linear, branched and cycloalkanes.

4. Write out the structural formulas of molecules for which the IUPAC names are given.

5. Memorize the prefixes and suffixes that are used in naming molecules and provide information about the carbon-carbon bond and the functional group.

6. Become proficient in drawing the conformations for cyclic hydrocarbons and in distinguishing between axial and equatorial positions in cycloalkanes.

7. Recognize molecules that can exhibit cis- and trans-isomers.

8. Characterize the most important chemical and physical properties of alkanes.

9. Outline the most important characteristics of sp^3 hybridization, which is the hybridization scheme used by carbons in alkanes.

10. Define the important terms and comparisons in this chapter and give specific examples where appropriate.

IMPORTANT TERMS AND COMPARISONS

Hydrocarbons
Saturated and Unsaturated Hydrocarbons
Aliphatic Hydrocarbons, C_nH_{2n+2}
IUPAC Systematic and Common Names
Molecular and Structural Formulas
Of Compounds
Straight Chain and Branched Chain
Cycloalkanes
Chair and Envelope Conformations
Cis- and Trans- Isomers
Physical Properties
Combustion or Oxidation of Alkanes
Chlorofluorocarbons (CFCs)
Petroleum
Octane Rating
Gasohol
Alkanes (-anes)
Line-Angle Formulas
Constitutional or Structural Isomers
Methylene Group ($-CH_2-$)
Alkyl Groups
Lewis Structures
Di-, Tri-, Tetra-, Penta- and Hexa-
Free Rotation and Conformations
Axial and Equatorial Bonds
Sterioisomers
Chemical Properties
Haloalkanes
Natural Gas
Fractional Distillation
Engine "Knock"

SELF-TEST QUESTIONS

MULTIPLE CHOICE. In the following exercises, select the correct answer from the choices listed. In some cases, more than one choice will be correct.

1. Consider the line-angle drawing shown below. This represents the molecule:

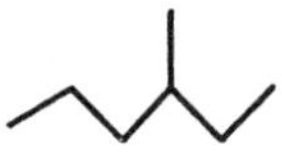

a. Heptane
b. 3-Methylhexane
c. 2-Methylhexane
d. 4-Methylhexane

2. Indicate which of the following molecular formulas represents a hydrocarbon.

a. C_3H_7Br
b. C_3H_6O
c. C_3H_8
d. C_3Br_8

3. Indicate which hydrocarbon has one methylene ($-CH_2$ unit) more than in $CH_3CH_2CH_2CH_2CH_2CH_2CH_3$.

a. butane
b. octane
c. propylene
d. heptane

4. Indicate which of the following is not a constitutional isomer for C_5H_{12}. Redraw these as a line-angle formula to help to distinguish them.

a. $CH_3CH_2CH_2CH_2CH_3$

b.
```
 CH3
  |
 HCCH2CH3
  |
 CH3
```

c.
```
    CH3
     |
CH3-C-CH3
     |
    CH3
```

d.
```
CH3CHCH2CH2CH3
    |
   CH3
```

5. Which of the following structural formulas represent the same compound?

```
      CH2-CH2-CH2-CH3                     CH3        CH2-CH3
       |                                   |          |
H3C-CH                 H3C-CH2-CH2CH2-CH             CH2
       |                                   |          |
      CH3                                 CH3        CH2-CH2-CH2
                                                              |
                                                             CH3
        (i)                  (ii)                      (iii)
```

a. all are different
b. (i) and (ii)
c. (i) and (iii)
d. all are equivalent

6. The IUPAC name for the compound below is:

```
H3C-CH2-CH2-CH2
               |
              CHCH2CH(CH3)2
               |
              CH2-CH3
```

a. 2-Methyl-4-propylheptane
b. 4-Ethyl-7-methylhocttane
c. 4- Ethyl-2-methyloctane
d. Methyl-decane

7. The structure for 5-Bromo-2-chloroheptane is:

a. 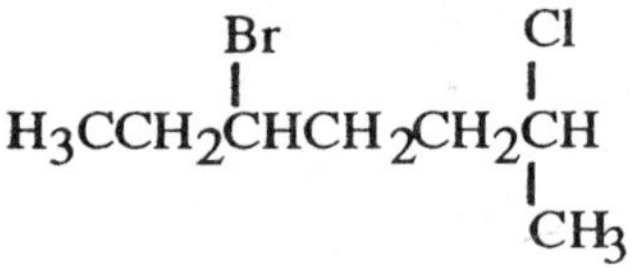

c. 

b. 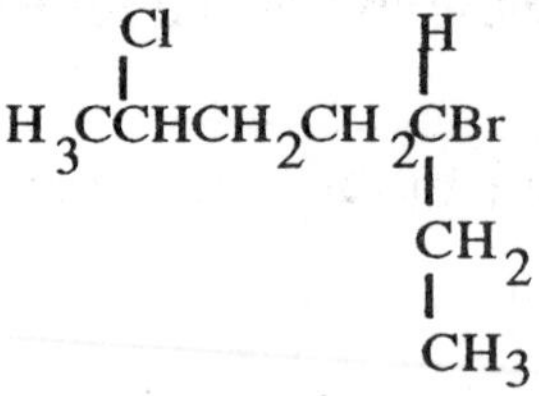

d. 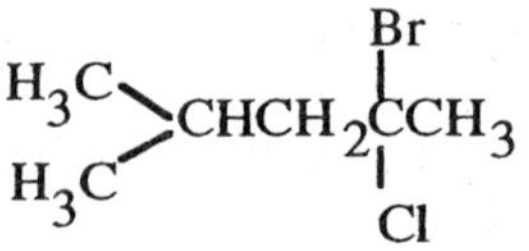

8. The combustion of pentane produces:

a. an alcohol
b. carbon dioxide and water
c. pentene
d. ethane

9. Indicate which of the following sets of molecules are constitutional isomers.

a. $CH_3CH_2OCH_3$ and $CH_3CH_2CH_2OH$
b. cyclohexane and $CH_2CHCH_2CH_2CH_2CH_3$
c. CH_3OH and CH_3CH_2OH
d. CH_3CH_2CHO and $CH_3CH_2CH_2OH$

10. The IUPAC name for 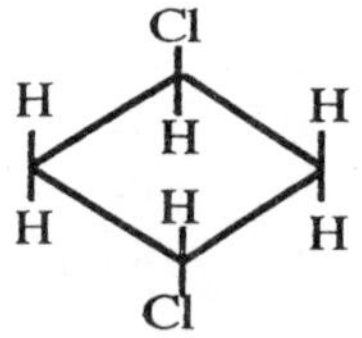 is:

a. trans-1,3-Dichlorocyclobutane
b. cis-1,2-Dichlorocyclobutane
c. cis-1,3-Dichlorocyclobutane
d. Cyclobutane

11. Indicate how many bromines in this brominated sugar are in the axial positions.

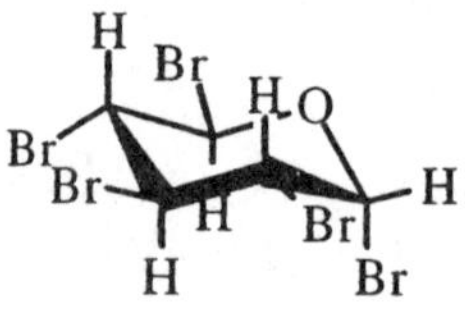

a. 1
b. 3
c. 2
d. 4
e. 5

12. The bonding in octane involves sp^3 hybrid orbitals on all carbon atoms. Indicate which of the following is true about the orbital and the bonding?

a. each hybrid orbital has a 50% contribution from the s and the p orbital
b. each hybrid orbital is a mixture of one part s orbital and three parts p orbital
c. the angle between the hybrid orbital is 109.5 degrees.
d. there are three sp^3 hybrid orbitals formed on each carbon

COMPLETION. Write the word, phrase or number in the blank space or draw the appropriate structure in answering the question.

1. Examine the following structural formulas and identify each of the molecules in one or more of the following categories listed.

(i) $CH_3CH_2CH_2CH_2CH_3$

(ii) $H_3COCH_2CH_3$

(iii) 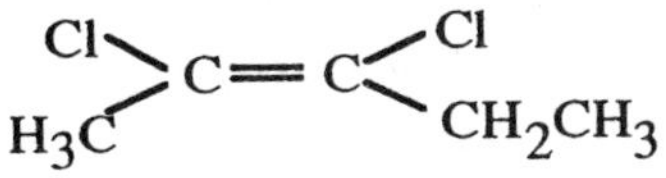

(iv) 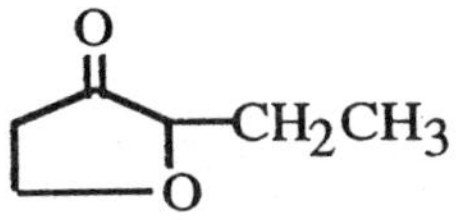

(v) 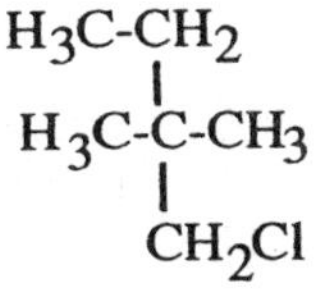

a. Hydrocarbon molecule ____________________
b. Contains aldehyde functional group ____________________
c. Halogenated hydrocarbon ____________________
d. Unsaturated (C=C) bond ____________________
e. Straight chain molecule ____________________
f. Contains more than one functional group ____________________
g. Acyclic molecule ____________________

2. The outstanding chemical characteristic of alkanes is their ____________________.

3. Isomers are defined as molecules which have the same ________________, but different ________________.

4. The ________ positions in the chair conformation of cyclohexane are found pointing up and down, respectively, on adjacent carbons.

5. A cis-isomer for 1,2-dimethylcyclopentane must have each of the methyl groups directed in the ______ direction on two __________ carbon atoms.

6. The boiling point of a series of alkanes increases as the molecular weight of the alkane ________________.

7. Cyclohexane in the chair form has __________ (a number) hydrogens "pointing" up or down called ________ hydrogens, in addition to the ______________ hydrogens "pointing" approximately in the "plane" of the ring called ________ hydrogens.

8. Cis- and trans-isomers are examples of __________ isomers.

9. Write the IUPAC name for the following molecules:

a. 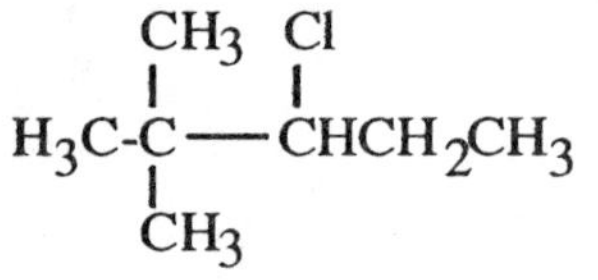 ______________________

b. 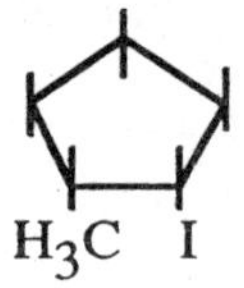 ______________________

c. 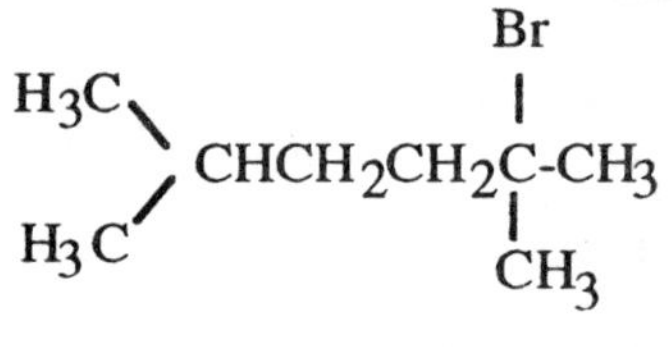 ______________________

10. Draw the structures for the following compounds, showing all the atoms.

    a. 3-Chloro-2-methylpentane ______________________

    b. 1-Iodo-3,3-dimethylcyclobutane ______________________

    c. 4-Isopropylheptane ______________________

11. A saturated hydrocarbon can exist in a number of __________ because there is __________ rotation about a (C-C) single bond.

12. Balance the following equation for the combustion or oxidation of an alkane.

$$C_5H_{12} \quad + \quad O_2 \quad = \quad CO_2 \quad + \quad H_2O$$

ANSWERS TO SELF-TEST QUESTIONS

<u>Multiple Choice</u>

1. b
2. c
3. b
4. d
5. b
6. c
7. a (and b) are the same molecule
8. b
9. a, b
10. a
11. a
12. b, c

<u>Completion</u>

1. a. i only; all other molecules contain other elements, in addition to C and H.
   b. none
   c. iii, v
   d. iii

e. i, ii
f. iv; iii also, if alkyl halide is considered as a functional group
g. i, ii, iii, v

2. lack of reactivity (except for burning or complete oxidation)
3. molecular formula, structural formula
4. axial
5. same, adjacent
6. increases
7. 6, axial, 6, equatorial
8. stereo or geometrical
9. a. 3-Chloro-2,2-dimethylpentane
   b. cis-2-Iodo-1-methylcyclopentane
   c. 2-Bromo-2,5-dimethylhexane

10.

a. 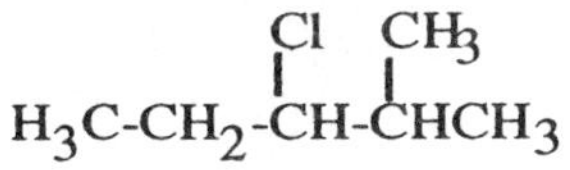

b. 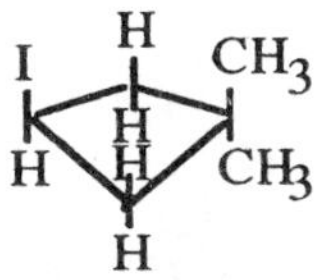

c. 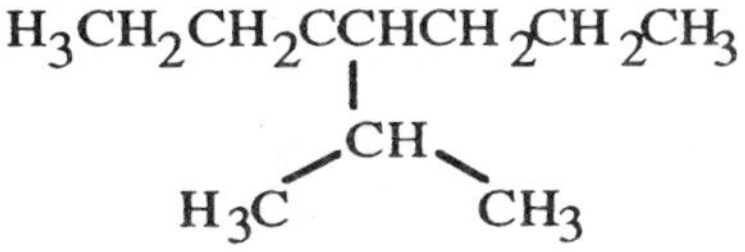

11. conformations, free
12. 1 C_5H_{12} + 8 O_2 = 5 CO_2 + 6 H_2O

On Chemistry:
It's hard and its hard, ain't it
hard, Good Lord
- Woody Guthrie
"Hard, Ain't It Hard," 1952

12

Alkenes

CHAPTER OBJECTIVES

Chapters 11 and 12 begin to present some of the substance of organic chemistry. This includes not only the formal naming of, but also the drawing of structures for organic molecules containing a particular functional group. In addition, there will be an increasing emphasis on recognizing the physical properties, and especially the chemical reactions, that can take place as a direct consequence of the presence of the functional group (i.e., the "reactive center"). The strategies and learning patterns developed in studying the material in this chapter will also be especially useful in subsequent chapters.

After you have studied the chapter and worked the assigned exercises in the text and the study guide, you should be able to do the following specific objectives.

1. Recognize the structural differences in alkenes, alkynes and distinguish them from aromatic (arene) molecules.

2. Given the structural formulas for both alkenes and alkynes, name the unsaturated molecules using the IUPAC rules.

3. Given the IUPAC name for an alkene or alkyne, draw the structural formula that shows all the atoms.

4. Characterize the addition reactions in terms of the proposed intermediates and the products that occur as H_2, HCl and Cl_2 react with an alkene or alkyne.

5. Show how the orbital hybridization of carbon in alkenes and alkynes determines the geometry about the carbon atom.

6. Use Markovnikov's rule, as it applies to the addition of an unsymmetrical small molecule (HX, X = Cl, Br, OH) to an alkene or alkyne, to predict the reaction products.

7. Outline the manner in which ethylene and substituted ethylenes undergoes chain-growth polymerization to form useful polymers.

8. Define the important terms and comparisons and give specific examples where appropriate.

IMPORTANT TERMS AND COMPARISONS

Unsaturated and Saturated Hydrocarbon
Alkane, Alkene and Alkyne
Monomer and Polymer
IUPAC Nomenclature
Cis-Trans Isomerization
Cycloalkenes
Carbocation
Oxonium Ion
Regioselective Reaction
Plastics

Reaction Types For Alkenes
a. Addition
b. Catalytic Hydrogenation
c. Hydration
d. Chain-Growth Polymerization
Markovnikov's Rule
Reaction Mechanism
Isoprene Units and Terpenes
Monomers and Polymers

Before challenging your comprehension of this material by working out the SELF-TEST, a few words are in order to help sort out the major themes presented. As this and subsequent chapters are individually addressed, it will become clear that although the general objectives and the nature of the material are similar, the strategy in learning is much the same. This is especially true in Chapters 13 through 18. Each of these chapters introduces one or more families of molecules with a characteristic functional group. **IT IS THE FUNCTIONAL GROUP THAT IS THE REACTIVE CENTER OF THE MOLECULE.** That is, this is where the reaction will take place with other molecules. The functional group will be the dominant factor in defining the chemical and physical properties of a compound. Compare the chapter titles with the list of functional groups. The focus of Chapter 12 is aimed at basically three major aspects of the two families of unsaturated molecules (alkenes, which contain one or more (C=C)double bonds, and alkynes, which contain a triple bond). The three themes are:

1. The **naming** of unsaturated molecules using the internationally accepted IUPAC rules. This nomenclature systematically defines the composition and structure of the molecule.

2. The **characterization of the physical properties** of each family of molecules.

3. **Identification of the reactions** that characterize each type of unsaturated molecule. These same general themes will reoccur in the subsequent chapters, with simply the focus reset on a different functional group.

Possibly the most troublesome area for students is developing a strategy in identifying and characterizing the reactions associated with a family of molecules. A guide or "road map" for the process, which is both simple and useful, is the following.

1. Identify the functional group.
2. Tabulate the reactions associated with the functional group.
3. Characterize the reaction by its general name (addition reaction, oxidation, substitution, etc.) and then tabulate any additional specifics (catalyst or heat required, follows Markovnikov's rule, etc.).

It is very helpful if the student can write out a condensed summary of the reactions for each family of molecules. An example for alkenes follows.

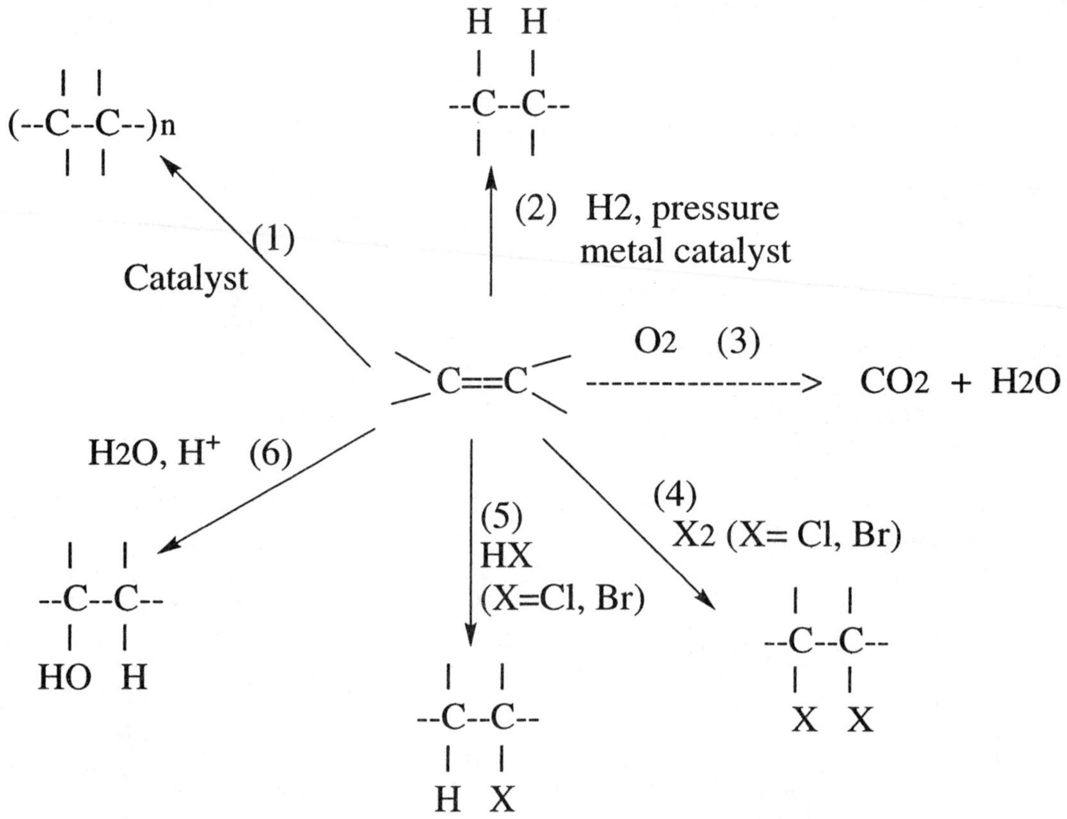

Condensed Reaction Summary for Alkenes

Reaction	**Name**	**Comment**
(1)	Polymerization	Requires appropriate catalyst; addition reaction
(2)	Catalytic hydrogenation	Addition reaction; pressure and metal catalyst
(3)	Combustion or oxidation of a hydrocarbon	CO_2 and H_2O are the common combustion products of all hydrocarbons
(4)	Bromination or chlorination	Addition reaction
(5)	Addition reaction of an HX gas	Obeys Markovnikov's rule
(6)	Hydration	Obeys Markovnikov's rule

SELF-TEST QUESTIONS

MULTIPLE CHOICE. In the following exercises, select the correct answer from the choices listed. In some cases, two or more answers will be correct.

1. A terpene containing 20 carbons has the following characteristics:

   a. 4 isoprene units
   b. 4 double bonds
   c. the product of a plant
   d. contain alkyne units

2. The reaction of cycloheptene with HCl yields

   a. 1, 2-Dichlorocycloheptene
   b. 2-Chloroheptane
   c. Chlorocycloheptane
   d. 1-Chlorocylcohexane

3. The carbocation intermediate formed in the reaction of HBr with 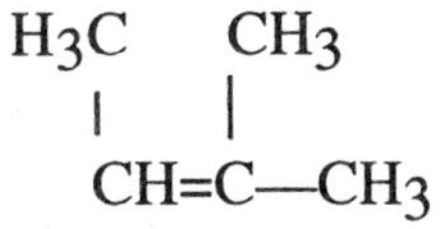 is:

   a. 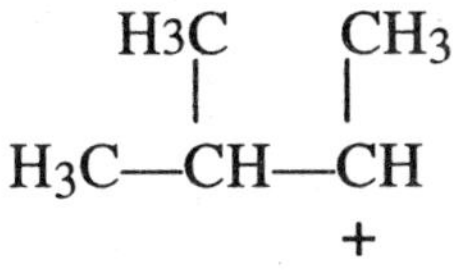

   c. 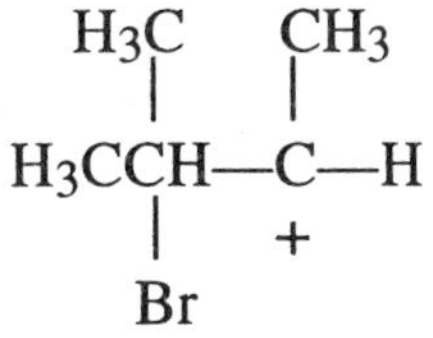

   b. 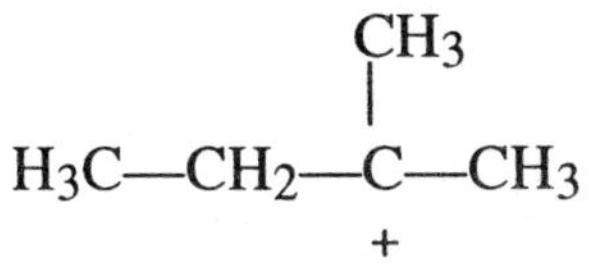

   d. 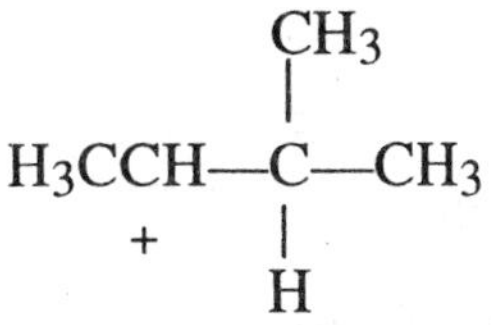

4. The principal product of the reaction of Br_2 with methylcyclopentene is:

   a. 2-Bromo-1-methylcyclopentane
   b. 1, 2-Dibromo-1-methylpentane
   c. no reaction
   d. 1, 2-Dibromo-1-methylcyclopentane

5. The product of the reaction of HBr with which of the starting materials listed below yields 2-Bromo-2-methylbutane.

   a. 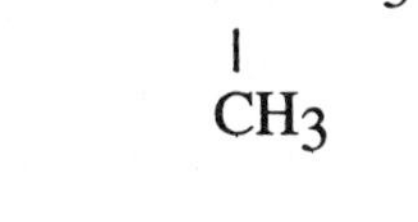

   c. 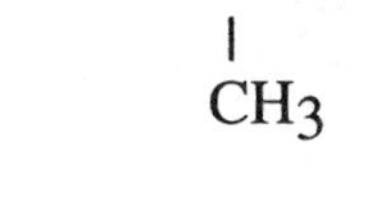

b. $CH_3CH_2CH_2CH_2CH_3$

d.

$$\begin{array}{c} CH_3 \\ | \\ CH_2CCHCH_2 \end{array}$$

6. The polymerization of vinyl chloride ($H_2C{=}CHCl$) produces:

   a. unsaturated polyvinylchloride
   b. saturated polyvinylchloride
   c. a branched polymer
   d. a linear polymer

7. An example of a secondary carbocation is:

   a. CH_3^+

   b. $H_3C{-}\overset{+}{C}H_2$

   c. $$\begin{array}{c} CH_3 \\ | \\ H_3C{-}C^+ \\ | \\ CH_3 \end{array}$$

   d. $H_3C{-}\overset{+}{C}H{-}CH_3$

8. The structural formula for the cis-isomer of $ClHC{=}CCH_3Cl$ is:

a.

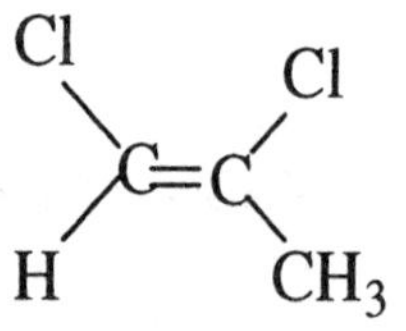

c.

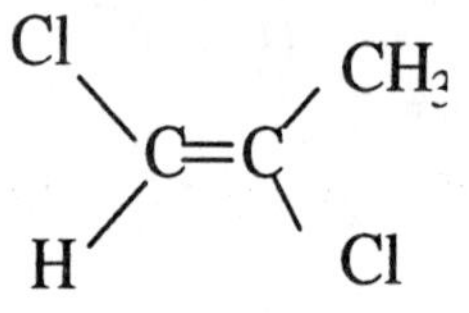

b.

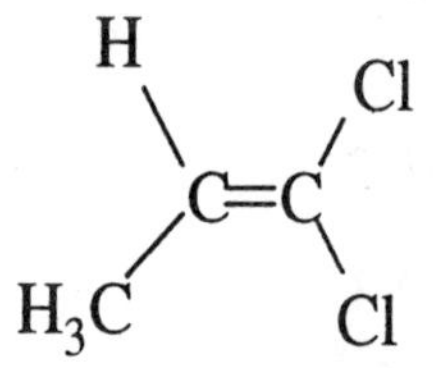

d.,

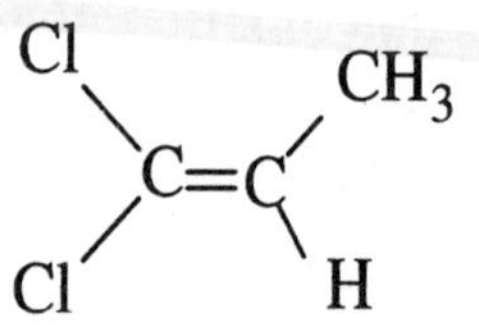

9. The reaction of HBr with which of the compounds shown below will produce $H_3CCH_2CHBrCH_3$:

a. $H_2C{=}CHCH_2CH_3$
b. $H_3CCH{=}CHCH_3$
c. $H_3CCH_2CH_2CH{=}CH_2$
d. $H_3CC{\equiv}CCH_3$

10. The IUPAC name for the compound below is:

$$\begin{array}{c} \quad\quad Br \\ \quad\quad | \\ H_3CCH_2CHCCH_2CH_3 \\ \quad\quad\;\; \| \\ \quad\quad\;\; CH_2 \end{array}$$

a. 3-Bromo-2-ethyl, 1-pentene
b. 3-Bromo-2-ethyl, 2-hexene
c. 3-Bromo-4-ethyl, 1-pentene
d. 3-Bromo-4-ethyl, 1-hexane

11. The name of this compound is:

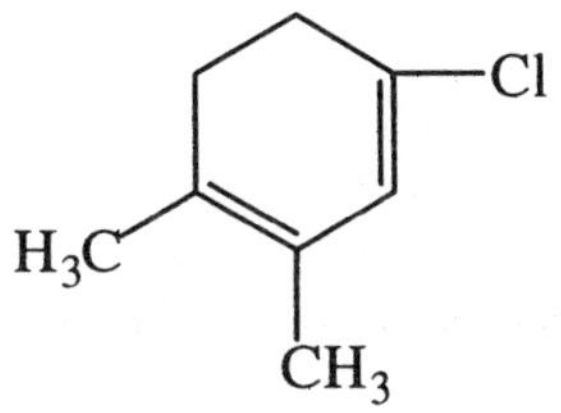

a. 1-Chloro-3, 4-dimethyl-2, 3-cyclohexatriene
b. 1-Chloro-3, 4-dimethyl-1, 3-cyclohexadiene
c. 3, 4-Dimethyl-1-chloro-1, 3-cyclohexadiene
d. 1-Chloroorthomethyl-1, 4-cyclohexadiene

12. Indicate which of the carbocations shown below is the most stable and the least stable.

a. $CH_3CH_2CH_2\overset{+}{CH_2}$

b. $CH_3CH\overset{+}{C}CH_3$ with CH_3 on the central CH: $\begin{array}{c} CH_3CHCCH_3 \\ | \\ CH_3 \end{array}$ (+ on C)

c. $CH_3\overset{+}{C}HCH_2CH_3$

d. $CH_3CH_2CH_2\overset{+}{C}H$ with CH_3: $\begin{array}{r} CH_3CH_2CH_2CH \\ | \\ CH_3 \end{array}$

COMPLETION. Write the word, phrase or number in the blank or draw the appropriate structure in answering the question.

1. Give the common or IUPAC names for the following molecules.

a.

```
       CH3
        |
CH2=CHCH=CH2
```

b. (-CH2CH2CH2)n

c. C_5H_8

d.

e.

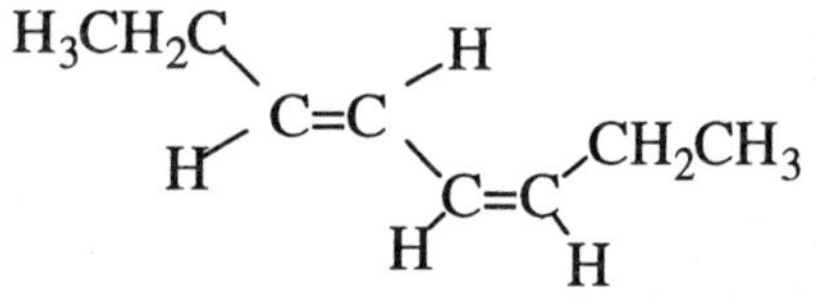

2. Fill in the Table below with the characteristics for alkanes, alkenes and alkynes.

Characteristic	Alkanes	Alkenes	Alkynes
Infix for Hydrocarbon type			
Bonding Angle (^{0})			
Relative (C-C) Bond Length			
Addition Reactions (yes or no)			

3. Draw the structures for the products of reactions b-d. In reaction a, only indicate the products.

a. CH3CHCH2 + O2 ⟶

b.

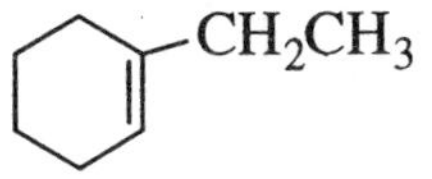

+ Br2 ⟶

c.

```
         CH3              H2SO4
        /
CH3CH=C          +  H2O  ⟶
        \
         CH3
```

d.

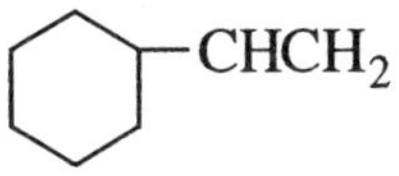

+ H2 —Pd⟶

4. Plastic combs are made of the polymeric material polymethylmethacrylate while indoor-outdoor carpeting is made of polypropylene. Draw the structural formulas for these products that are produced in the reactions below.

a.

```
        O                  O
        ||                 ||              catalyst
H2C=C-C-OCH3  +  H2C-C-C-OCH3  ⟶  ______________
```

b.

```
                                catalyst
H2C=CHCH3  +  H2C=CHCH3  ⟶  ______________
```

c. This polymerization reaction is also classified as an ________________ reaction.

d. This is an example of a copolymer because the polymer is made from ______________.

5. The boiling point for 1-hexene is ___________ than that of 1-octene.

6. The unsaturated molecule, CH_3CCH, methylacetylene, can be hydrogenated, similar to that discussed for ethylene. The structure for the product of this hydrogenation is ______________. This product can then be reacted with HCl to produce a saturated alkylchloride. The structure for this final product is ____________ .

7. Markovnikov's rule states ____.

8. Draw the structural formula for 1, 2, 3,-Tribromo-4-chlorobutadiene.

9. Alkynes undergo the very same types of reactions as alkenes. With this in mind, draw the products for the sequence of the two reactions shown below:

$HC\equiv CCH_3$ + Cl_2 ———> ____________________
product 1

____________________ + H_2O ———> ____________________
product 1

10. Squalene is a terpene compound involved in the biosynthesis of cholesterol in the body. The condensed structure is shown below and can be regarded as a polymer of isoprene monomer units.

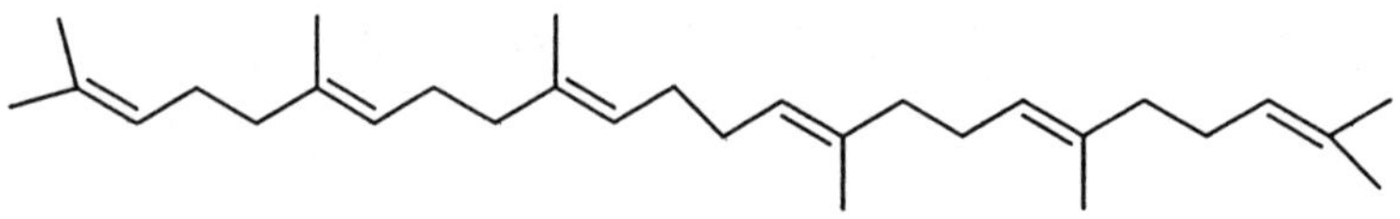

a. The structural formula for an isoprene unit is ______________________________.

b. Although the many structures of terpenes are very diverse, they all share two common characteristics.

(i) ______________________________

(ii) ______________________________

c. Determine the number of isoprene units in squalene and underline these units in the structure shown.

number of isoprene units = ____________________

ANSWERS TO SELF-TEST QUESTIONS

Multiple Choice

1. a, b, c
2. c
3. b
4. d
5. c
6. b, d
7. d
8. a
9. a
10. a
11. b
12. b, most stable; a, least stable

Completion

1. a. Isoprene or 2-Methyl-1, 3-butadiene
   b. Polyethylene

c. Cyclopentene
d. 1, 4-Cyclohexadiene
e. trans, cis-3, 5-Octadiene

2.

Characteristic	Alkanes	Alkene	Alkynes
Infix for Class Of Hydrocarbon	-an-	-en-	-yn-
Bonding Angle (°)	109.5	120	180
Relative (C-C) Bond Length	longest	intermediate	shortest
Addition Reactions	no	yes	yes

3.

a. CH3CHCH2 + O2 ⟶ CO2 + H2O

b.

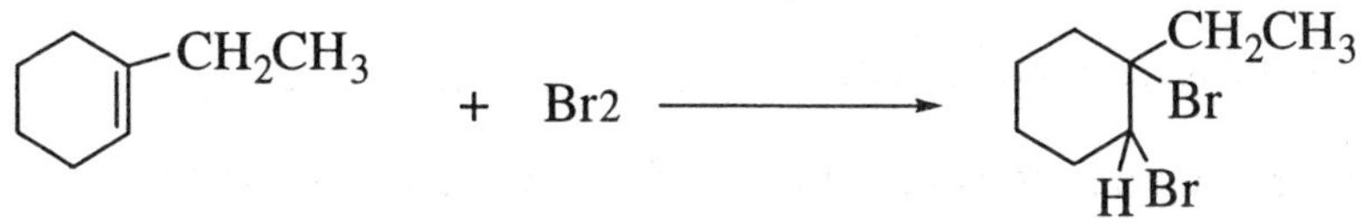

c.

```
         CH3                H2SO4          CH3
        /                                 /
CH3CH=C        +   H2O   --------->  CH3CH2-C-OH
        \                                 \
         CH3                               CH3
```

d.

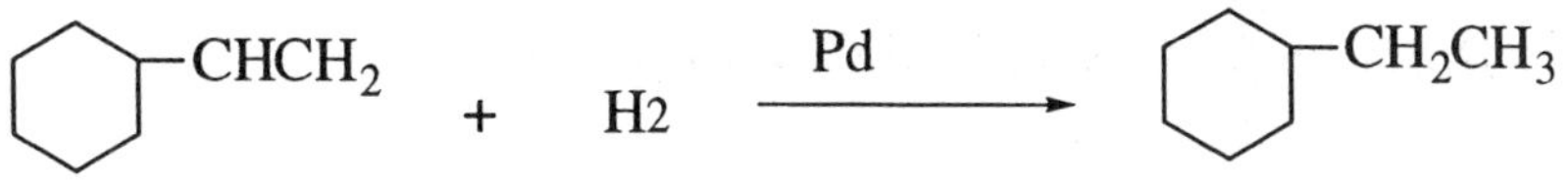

4. The answers in parts (a) and (b) show the repeating unit in each polymer.

a.

```
     O       O
     ||      ||
  H3COC     COCH3
     |       |
-(-CH2-C-CH2-CH-)n-
       |
      CH3
```

b.

```
      CH3
       |
-(-CH2-CH-)n
```

c. addition

d. two different monomer units

5. lower

6.

```
CH3 H          CH3 H
 |   |          |   |
 C = C ;      H-C---C-Cl
 |   |          |   |
 H   H          H   H
```

7. When an unsymmetrical reagent such as HX adds to an alkene or alkyne, the hydrogen from HX adds to the carbon atom of the double bond which already has the greater number of hydrogen atoms.

8.

```
Cl  Br  Br  Br
 |   |   |   |
HC = C - C = CH
```

9. Product 1

```
  Cl    Cl
  |     |
H-C  =  C-CH3
```

Product 2

```
  Cl  Cl
  |   |
H-C - C -CH3
  |   |
  H   OH
```

10.

```
       CH3
       |
a  H2C=C-C=CH2
         |
         H
```

b. (i) They are synthesized in plants and are therefore called "natural products".
(ii) All follow the isoprene rule.

c. 6 isoprene units

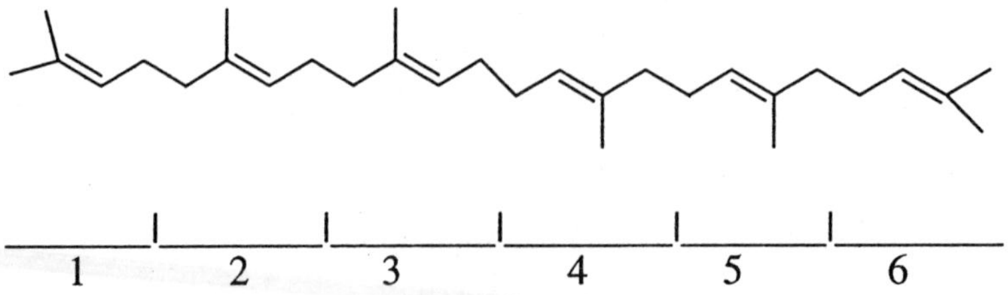

A great teacher is one who makes
himself progressively unnecessary.
- Thomas Carruthers

13

Benzene and Its Derivatives

CHAPTER OBJECTIVES

After you have studied the chapter and worked the assigned exercises in the text and the study guide, you should be able to do the following.

1. Recognize aromatic compounds and aromatic groups in larger, more extended structures.

2. Draw the Lewis structure of the aromatic molecule, benzene, in its individual resonance forms and in its resonance hybrid form.

3, Given the structural formulas for a variety of aromatic molecules, name the molecule using the IUPAC rules of nomenclature.

4. Given the IUPAC names for aromatic compounds, draw their structures, showing all atoms.

5. Characterize the type of chemistry that aromatic molecules exhibit, including substitution reactions involving halogenation, nitration and sulfonation. Note that these reactions require a catalyst.

6. Write out the acid base reaction for a phenol titrated with NaOH.

7. Contrast the chemical reactions associated with alkenes with those for aromatic molecules.

8. Define the important terms and comparisons in this chapter and give specific examples where appropriate.

IMPORTANT TERMS AND COMPARISONS

Aromatic or Arene Compounds
Resonance and Aromatic Sextets
Ortho, Meta and Para
Polynuclear Aromatic Hydrocarbons (PAHs)
Addition Versus Substitution Reactions
Phenol and the Phenyl Group
Acid/Base Chemistry of Phenol
DDT
Antioxidants, Vitamin E, BHT and BHA

Kekule Structure
Aryl, Phenyl, and Benzyl Groups
Carcinogen
TNT (2, 4, 5-Trinitrotoluene
Halogenation, Nitration and Sulfonation of Aromatic Molecules
Ka and pKa Values and Acidity
Radical

SELF-TEST QUESTIONS

MULTIPLE CHOICE. In the following exercises, select the correct answer from the choices listed. In some cases, two or more answers may be correct.

1. Indicate which of the following properties is a characteristic of benzene and its derivatives.

   a. Soluble in organic solvents and insoluble in water
   b. Burns in air to form carbon dioxide and water
   c. Bonding is essentially covalent
   d. Reacts with acids and bases

2. The nitration, halogenation and sulfonation reaction of benzene is generally called a

   a substitution reaction
   b. addition reaction
   c. elimination reaction
   d. dehydration reaction

3. The name of the following molecule is

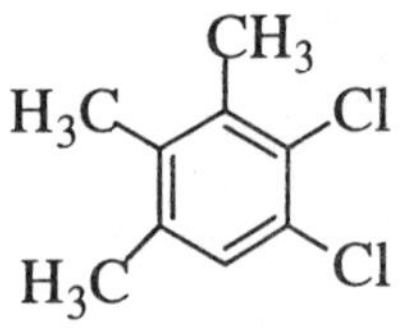

   a. 1, 2-Dichloro-4, 5-dimethy-3-toluene
   b. 5, 6-Dichloro-2, 3-dimethyltoluene
   c. 2, 3-Dichloro-4, 5, 6-trimethyltoluene
   d. 1, 2, 3-Trimethyl-5, 6-dichlorobenzene

4. The individual (C-C) bonds in each Lewis structure for benzene are either single or double bonds. The (C-C) bond length in benzene is actually

   a, alternating single and double bonds
   b. a bond type midway between a single and double bond
   c. primarily single bonds
   d. primarily double bonds

5. In polynuclear aromatic hydrocarbons (PAHs), the aromatic rings

   a. share a single carbon atom
   b. always share three or more carbon atoms
   c. are held together by a single linking carbon atom
   d. share one or more sides of the aromatic rings

6. Reaction of I_2 with the aromatic ring in the protein, thyroglobulin, starts the synthesis of thyroxine. Indicate what type of reaction this is.

   a. addition reaction
   b. elimination
   c. substitution
   d. oxidation

7. In a disubstituted benzene, the greatest possibility for the substituted groups "bumping into" each other is when the two groups are in

a. ortho positions
b. para positions
c. meta positions
d. either ortho, meta or para

8. Phenol exhibit a significant acidic character, much more than found in alcohols. The pKa for phenol is 9.95. At pH 9.95, the relative ratio of phenol/phenoxide is

a. 50/50
b. 30/70
c. 100% phenol
d. 100% phonoxide

COMPLETION. Write the word, phrase or number in the blank or draw the appropriate structure in answering the question.

1. Draw the structures for these reoccurring aromatic units.

a. phenol ________

b. aniline ________

c benzaldehyde ________

d. benzoic acid ________

e. anisole ________

f. toluene ________

2. Draw the structures for these reoccurring groups or substituents.

a. phenyl ________

b. benzyl ________

c. phenoxide ________

3. Name the three different isomers that are possible for a benzene ring that has one nitro- and one chloro- substituent in place of the hydrogens. Indicate the site numbers for these arrangements.

	Names	Numbers for the substitution sites
a	____________________	_______
b.	____________________	_______
c.	____________________	_______

4. Write out the complete equations for these reactions.

   a. chlorination of benzene

   b. sulfonation of benzene, followed by treatment with NaOH

   c. nitration of benzene

   d. titration of phenol with NaOH

   e. conversion of nitrobenzene to aniline

5. Draw the following molecules.

   a. Dibenzyliodomethane

   b. 4-Bromo-2-chlorobenzaldehyde

   c. p-Methylnitrobenzene (or p-Nitrotoluene)

   d. 4-Bromo-1, 2-Dichloro-3-toluene

6. Foods can undergo a reaction called __________ if they have carbon-carbon double bonds and are exposed to other reagents and _________ gas. These reactions are initiated by either _______ or light, which converts the unsaturated molecule to a _______. The unpaired electron resides in the molecule on the _______ atom. This leads to a chain ___________ reaction, which can lead to spoilage. A common defense is to add a chemical __________ such as ______ or ______. Nature's "scavenger" is vitamin ___.

ANSWERS TO SELF-TEST QUESTIONS

Multiple Choice

1. a, b, c (phenol is a weak acid, so add d for this special case)
2. a
3. a (the parent is toluene)
4. b
5. d
6. c
7. a
8. a

Completion

1,

a. 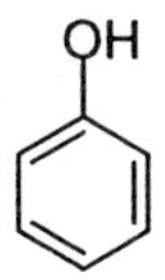

b. 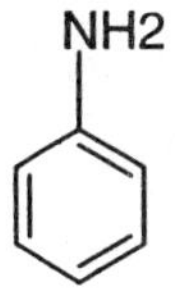

c. 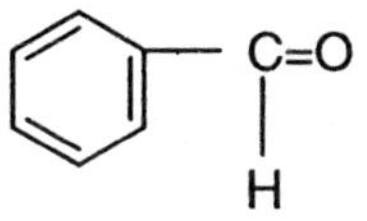

d. 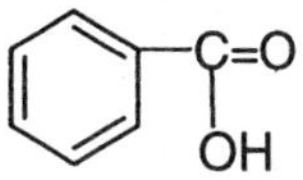

e. 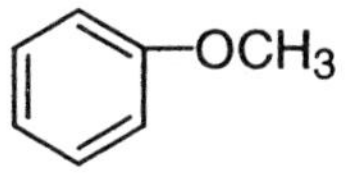

f. 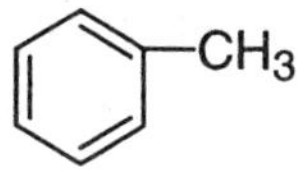

2.

a. 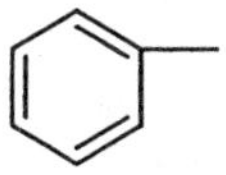

b. 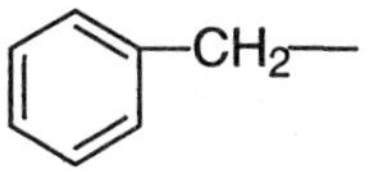

c. 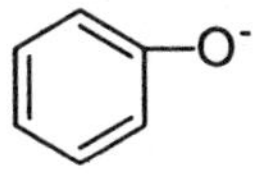

3.

a.	o-Chloronitrobenzene	sites 1 & 2
b.	m-Chloronitrobenzene	sites 1 & 3
c.	p-Chloronitrobenzene	sites 1 & 4

4.

a.

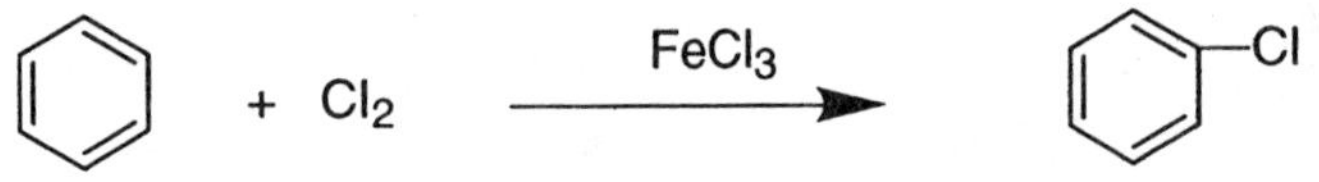

b.

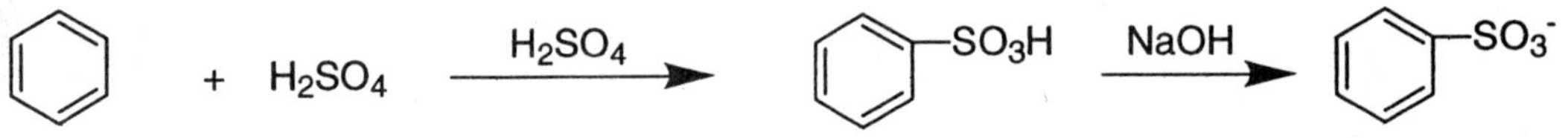

c.

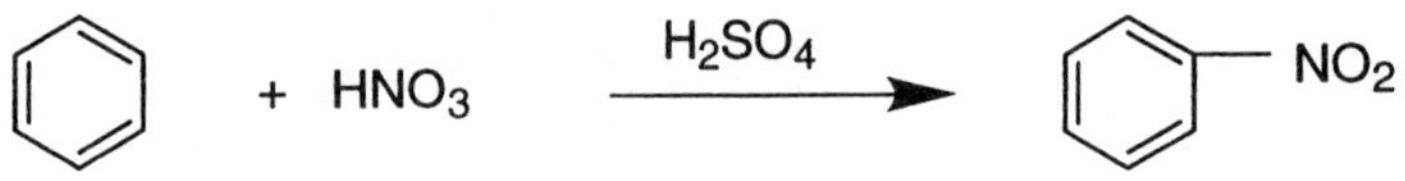

d.

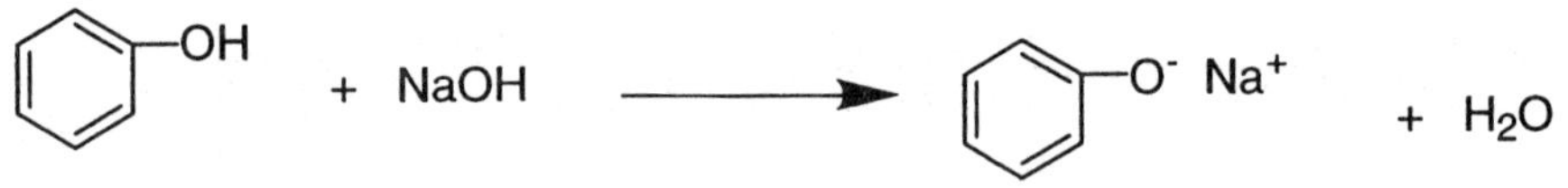

e.

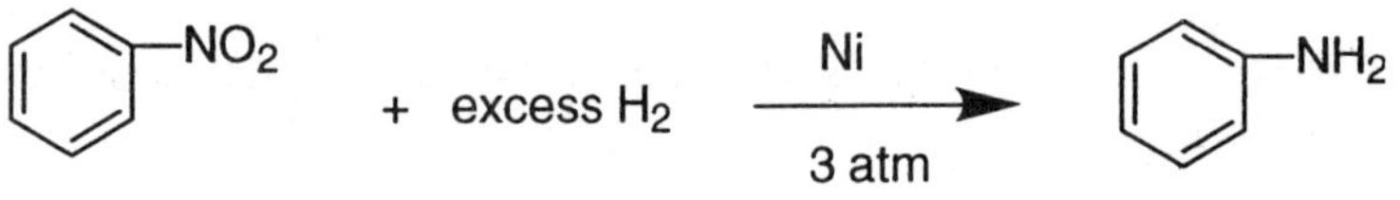

5.

a.

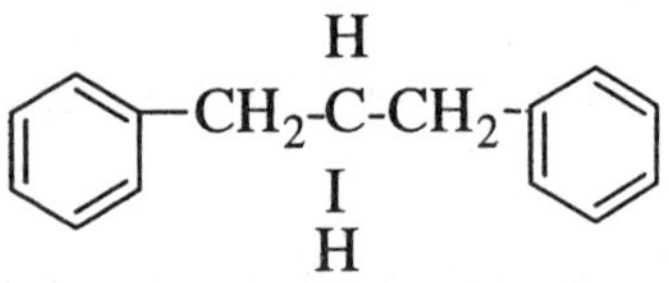

b.

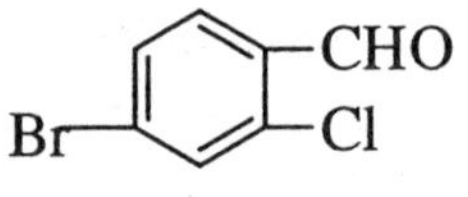

c.

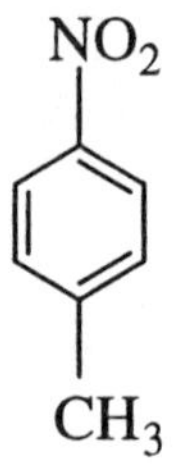

d.

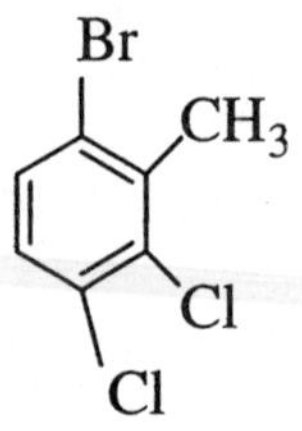

6. autooxidation, O2, heat, radical, carbon, elongation, BHT, BHA, E

More wine, less wits, you know; wine makes a man sing even if he is a rare scholar, makes him titter and chuckle, aye, makes him dance a jig, and makes him blurt out what were better kept to himself.

- Odyssey, XIV, 464

14

Alcohols, Ethers, and Thiols

CHAPTER OBJECTIVES

After you have studied the chapter and worked the assigned exercises in the text and the study guide, you should be able to do the following.

1. Identify compounds that contain an alcohol (R-OH), ether (R-O-R') or thiol (R-SH)functional group.
2. Given the IUPAC names for simple alcohols, ethers and thiols, draw the structural formula for each that includes all the atoms.
3. Given the structural formula for an alcohol, ether or thiol, name the compound according to the IUPAC rules.
4. Write out the reactions for the dehydration of alcohols and point out the predominant product in cases in which there are isomeric alkene products.
5. Draw the structures for primary, secondary and tertiary alcohols.
6. Write out the reactions for the oxidation of primary, secondary and tertiary alcohols and characterize the possible products formed.
7. Review the concept of hydrogen bonding and discuss the role of hydrogen bonding in understanding the boiling points of alcohols relative to hydrocarbons and solubility characteristics of alcohols and ethers in water.
8. Characterize the similarities and differences in the physical properties of alcohols, ethers, thiols and saturated hydrocarbons of comparable molecular weight.

9. Draw the structural formulas for thiols and disulfides and write out the oxidation/reduction reaction associated with them.

10. Prepare a summary sheet for the reactions of alcohols, ethers, thiols and disulfides.

11. Define the important terms and comparisons in this chapter and give specific examples where appropriate.

IMPORTANT TERMS AND COMPARISONS

Functional Group
Primary, Secondary and Tertiary Alcohols
Diol, Glycols and Triol
Alfred Nobel
Dehydration Reaction
Oxidation Reaction
London Dispersion Forces
Ethanol
Carboxylic Acid
Hydroxyl- and Sulfhydryl- Groups
Anesthetic
Alcohol, Ether and Thiol
IUPAC Nomenclature
Glycerol or Glycerin
Nitroglycerine and Dynamite
Elimination Reaction
Hydrogen Bonding
Hydrophobic
Aldehyde and Ketones
Thiols, Thioethers and Disulfides
Disulfides

SELF-TEST QUESTIONS

MULTIPLE CHOICE. Select the correct answer or answers from the choices listed. In some cases, more than one answer is correct.

```
                                OH
                                |
1. The chemical oxidation of H3CCH2CHCH3 produces:
```

1. The chemical oxidation of $H_3CCH_2CH(OH)CH_3$ produces:

```
          O                              O
          ||                             ||
a.  CH3CH2CCH3               c.   CH3CH2CH
```

b. $CH_3CH_2OCH_2CH_3$

d. CO_2 + H_2O

2. The compound that is most capable of exhibiting hydrogen bonding is:

a. $H_3COCH_2CH_3$
b. $CH_3CH_2CH_2CH_2OH$
c. butane
d. $CH_3CH_2S\text{-}SCH_3$

3. Indicate the type of chemical change that occurs in the alcohol during the breath test that law enforcement officers use to determine alcohol levels in drivers.

a. oxidation
b. dehydration
c. reduction
d. substitution

4. Indicate two characteristics that make ethers potentially dangerous.

   a. solubility in water
   b. flammability
   c. potential for explosion
   d. toxicity

5. The compounds that are widely recognized for their unpleasant odors are:

   a. thiols
   b. alcohols
   c. ethers
   d. alkanes

6. Tertiary alcohols are resistant to oxidation because they:

   a. are already oxidized
   b. do not have a H on the C that is bonded to the alcohol group
   c. are unsaturated
   d. readily form an alkene

7. The complete oxidation of a primary alcohol produces:

   a. ketone
   b. aldehyde
   c. carboxylic acid
   d. cyclic ether

8. The oxidation of CH_3CH_2SH results in a:

   a. thiol
   b. disulfide (bond formation)
   c. cyclic thiol
   d. useful anesthetic

9. The order of the compounds that increasingly take part in hydrogen bonding with water is:

   a. alkane < alcohol < ether < aldehyde
   b. alcohol < alkane < aldehyde < ether
   c. alkane < ether < aldehyde < alcohol
   d. ether < aldehyde < alcohol < alkane

10. Oxidation of the following alcohol by potassium dichromate in sulfuric acid ($K_2Cr_2O_7/H_2SO_4$) yields

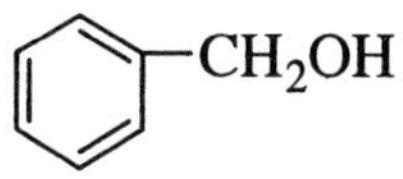

a.

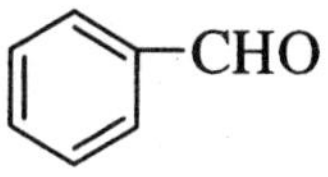

b.

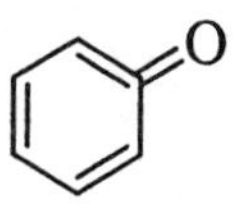

c.

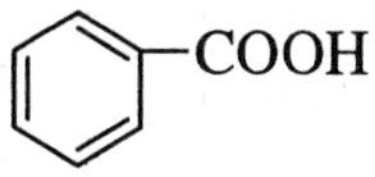

d.

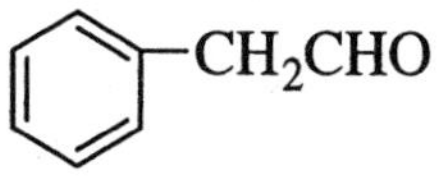

11. The oxidation of ethanol produces acetaldehye (CH_3CHO). Further oxidation of acetaldehyde yields

a. $CHOOH$
b. CH_3OH
c. CH_3COOH
d. CH_3COCH_3

12. Considering the potential for hydrogen bonding, predict what the relative boiling points for propanethiol and ethylmethylsulfide would be.

a. comparable
b. ethylmethylsulfide is higher
c. propanethiol is higher
d. cannot predict

13. Order the following compounds in terms of increasing acid character.

a. alcohol > thiol > alkane > water
b. thiol > alcohol ~ water > alkane
c. water > alcohol > alkane > thiol
d. thiol > alkane > water > alcohol

COMPLETION. Write the word, phrase or number in the blank or draw the appropriate structure in answering the question.

1. In the dehydration of an alcohol, two groups are removed from adjacent carbons to produce a double bond. This reaction often is referred to as an __________ reaction.

2. Tertiary alcohols cannot be _____________, while the oxidation of a primary alcohol yields either an ________________ or a ________________.

3. The relative order of boiling points for a simple alcohol and an ether, both having comparable molecular weights is:

_______________ > _______________

4. The relative solubility of the propanol, propane, 1,2-propanediol and ethylmethylether in water is:

_______________ > _______________ > _______________ > _______________

5. Complete the equations by writing the appropriate structural formula.

a.
H_3C CH_2OH
| |
$H_3CCHCHCH_3$ --------->(H_2SO_4) _________________________
initial product

_________________ ---------->([O]) _________________________
initial product final product

b. _________________ ---------->(H_2SO_4, 180^0 C) H_3CCH_2-O-CH_2CH_3

c. ________________ $\xrightarrow[\text{[H]}]{\text{reduction}}$ $HSCH_2CH(CH_3)CH_2SH$

```
                         reduction          CH3
                            [H]              |
c.   ________________   ---------->    HSCH2CHCH2SH
```

d.

```
        CH2CH3
          |              H2SO4
     CH3CH-OH      ----------------->
          |               heat
        CH3
```

e.

```
        CH2CH3
          |              K2Cr2O7
     CH3CH-OH      ------------------>
                         H2SO4
```

f. CH_3CH_2OH $\xrightarrow[180^0]{H_2SO_4}$

g. $CH_3CH_2\text{-}O\text{-}CH_2CH_3$ $\xrightarrow[\text{heat}]{H_2SO_4}$

h. $CH_3CH_2\text{-}SH$ $\xrightarrow{[O]}$

i.

```
                                                   O
                      K2Cr2O7                      ||
i.                --------------->    CH3CH2CHCCH3
                        heat
```

j.

```
     CH2–CH–CH2  +  3 HNO3    ---------->
      |    |    |
     OH   OH   OH
```

6. Characterize each of the following molecules as alcohols, ethers, both or none of these.

a. H3CCHO ____________________

b. H3CCH2CHCH2 ____________________

```
          \ /
           O
```

c. HOCH2CH2OH ____________________

d. ____________________

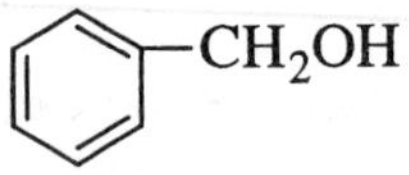

e. ____________________

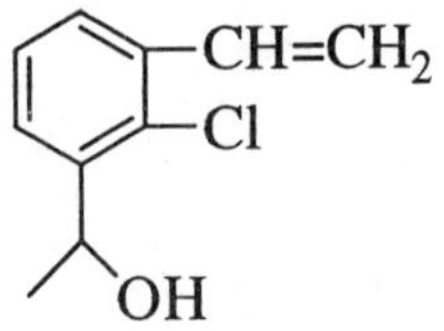

f. ____________________

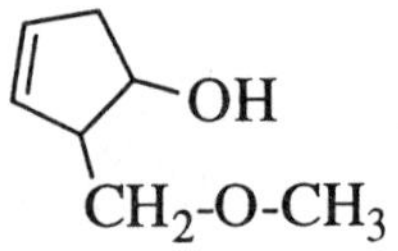

7. Give suitable names for the following compounds.

a. HOCH2CHOHCH2CH2OH

b. CH3CH2-O-CH(CH3)2

c.

```
     H3C OH
       |  |
H3C-C-CHCH3
       |
      CH3
```

d.

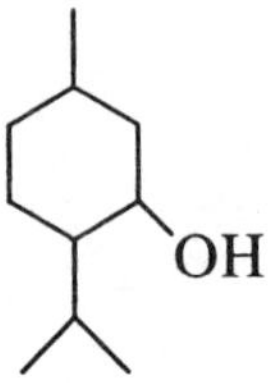

e. $H_3CCH_2CH_2SH$

8. Write out the structural formulas for the following compounds:

   a. Butylmethylether

   b. 4-Ethyl-3-hepten-1-ol

   c. 1, 2-Pentanediol

   d. Ethylene oxide

   e. Ethylene glycol, a common liquid used as an antifreeze in a car radiator.

9. In the reaction of potassium dichromate with ethanol, the dichromate is an _________ agent and becomes reduced in the reaction, while the alcohol is the ______ agent and becomes ________ in the reaction.

10. The cyclic ether, ethylene oxide, is especially reactive because ____________________.

ANSWERS TO SELF-TEST QUESTIONS

Multiple Choice

1. a
2. b
3. a
4. b, c
5. a
6. b
7. c
8. b
9. c
10. a and/or c
11. c
12. b
13. b

Completion

1. elimination
2. oxidized, aldehyde, carboxylic acid
3. alcohol > ether
4. 1, 2-propanediol > propanol > ethylmethylether > propane
5.

a.

```
CH3CHCHCH3,            CH3CHCHCH3
    |  |                   |  |
  H3C  C-H               H3C  C-OH
       ||                     ||
       O                      O
initial product        final product
```

b. CH_3CH_2OH

c.

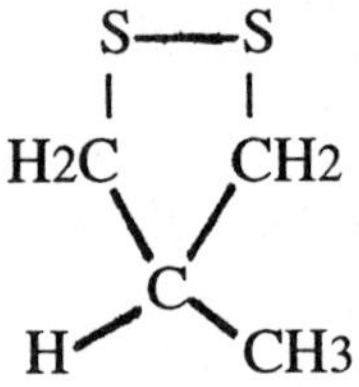

d.

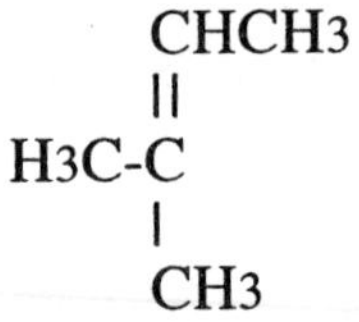

e.

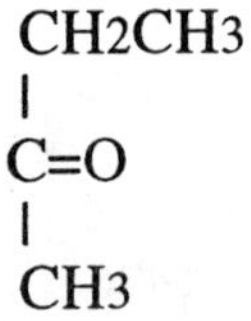

f. $CH_2{=}CH_2$

g. No reaction

h. $CH_3CH_2\text{-}S\text{-}S\text{-}CH_2CH_3$

i.

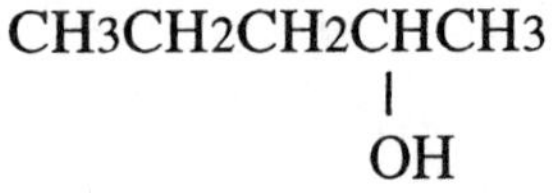

j.

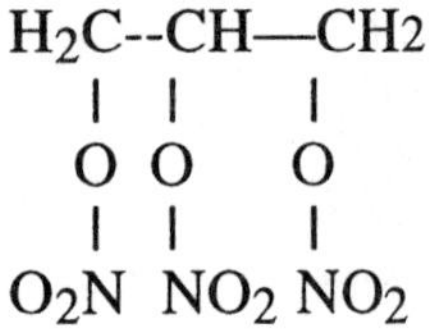

+ 3 H_2O

6. a. none
   b. ether
   c. alcohol
   d. alcohol
   e. alcohol
   f. both an alcohol and an ether

7. a. 1, 2, 4-Butanetriol
   b. Ethylisopropylether
   c. 3, 3-Dimethyl-2-butanol
   d. Menthol or 2-Isopropyl-5-methylcyclohexanol
   e. Propanethiol

8. a. $CH_3CH_2CH_2CH_2\text{-}O\text{-}CH_3$

b. $CH_3CH_2CH_2C{=}CHCH_2CH_2\text{-}OH$
|
CH_2CH_3

c. $H_2C\text{-}CHCH_2CH_2CH_3$
| |
HO OH

d. Ethylene oxide

e. Ethylene glycol (1, 2 Ethanediol)

CH_2—CH_2
| |
HO OH

9. oxidizing, reducing, oxidized

10. of the "bond strain" in the three-membered ring.

Imagination is more important than knowledge
- Albert Einstein

15

Chirality: The Handedness of Molecules

CHAPTER OBJECTIVES

After you have studied the chapter and worked the assigned exercises in the text and the study guide, you should be able to do the following.

1. Distinguish the different types of isomers that have been presented in this and previous chapters and indicate an example of each.
2. Describe how connectivity (of the atoms) distinguishes stereoisomers from constitutional isomers.
3. Determine whether a molecule exhibits a stereocenter.
4. Determine the number of stereocenters in a chiral molecule and the maximum number of stereoisomers, by using the 2^n rule.
5. Distinguish between enantiomers and diastereomers.
6. Indicate the relationship between a racemic mixture and a pair of enantiomers.
7. Become proficient in using the R, S system to assign an R or S configuration at each stereocenter.
8. Explain how plane-polarized light and a polarimeter are used to experimentally characterize an enantiomer.
9. Distinguish between the terms optically active, dextrorotatory, levorotatory and specific rotation.
10. Explain the importance of enantiomers in stereospecific interactions with enzymes in biological reactions.
11. Define the important terms and comparisons in this chapter and give specific examples where appropriate.

IMPORTANT TERMS AND COMPARISONS

Stereoisomerism
Stereocenter
Chiral and Achiral Molecules
Diastereomers
R, S Priorities For Groups
Chiral Drugs
Optically Active Molecule
Dextrorotatory (Clockwise; to the right)

Enantiomer and Enantiomeric Pair
Nonsuperposable Mirror Image
Racemic Mixture
R and S Configuration
2^n (Maximum of Stereoisomer) Rule
Plane Polarized Light and a Polarimeter
Specific Rotation
Levorotatory (Counterclockwise; to the left)

FOCUSED REVIEW

Before we finish reviewing the major functional groups in organic chemistry and embark into the subject of **biochemistry**, the important concept of chirality is presented. It will become increasingly clear that this concept, and the other chemical concepts presented in previous chapters will provide the foundation for an understanding of biochemistry. As you encounter these basic concepts, do not be shy about returning to these chapters to review and reclaim a clear understanding of any vague terms. For example, a few of the concepts and reactions that were presented in previous chapters and are relevant to this chapter are presented below. You may want to extend this summary, and prepare a similar review after each subsequent chapter.

Types of Isomers

Recall that **constitutional isomers** are molecules that have the same molecular formula, but different structural formulas (i. e., different molecular structures). As a result of the structural differences, the chemical and physical properties of the isomers are different. A simple example of this is the molecular formula, C_2H_6O, and the different molecules (isomers) with this same formula. This formula may represent either (a) dimethyl ether or (b) ethanol.

(a) H_3COCH_3 (b) H_3CCH_2OH

These two isomers have different **structures**, different **types of bonds**, different **functional groups** and correspondingly different chemical and physical properties. These isomers are often referred to as **structural isomers** or functional isomers.

Although structural isomers are in many cases very obvious, other isomer types are more subtle. **Stereoisomers** are compounds which have the same connections between atoms (i. e., each isomer has the same number and types of bonds), but have a different three-dimensional arrangements in space. There are two major classes of stereoisomers of interest to us.

The first type is the **geometric isomer** that results from the restriction of rotation about a bond. The most obvious type of isomers in this case may be the **cis- and trans-isomers** of a substituted olefin, such as $C_2Cl_2H_2$.

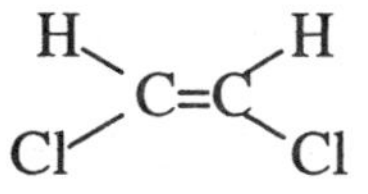

Cis-isomer

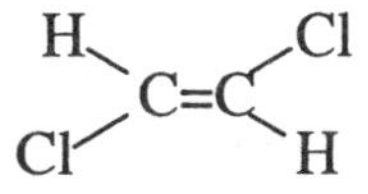

Trans-isomer

The second type is the **optical isomer,** which contains a stereocenter. Plane polarized light in a polarimeter is rotated on passing through a solution of an optical isomer. Two subdivisions of optical isomers are (a) enantiomers and (b) diastereomers.

In summary, the isomeric forms of molecules can be subdivided into groups as shown below. Much of our interest in this chapter will focus on simple organic molecules that are enantiomers or diastereomers. However, our interest in living cells will focus interest on biological molecules, including monosaccharides, amino acids, nucleotides, lipids and the major macromolecules such as polysaccharides, proteins and nucleic acids - all of which are chiral molecules and exhibit optical active.

Isomer Classification

Isomer Type	Character	Examples
Constitutional	Same molecular formula, but different molecular structure	H_3CCH_2OH H_3COCH_3
Stereoisomers	Same connectivity of atoms, but different three-dimensional arrangement in space	
a. Geometric	cis- or trans-arrangement about a carbon-carbon double bond, such as observed in some fatty acids (Chapter 120	cis- and trans-isomer of $C_2Cl_2H_2$
b. Optical		
(i) Enantiomers	Nonsuperimposable mirror images	D- and L-forms of glyceraldehyde
(ii) Diastereomers	Optical isomers which are not enantiomers	α- and β-forms of D-glucose

SELF-TEST QUESTIONS

MULTIPLE CHOICE. Select the correct answer from the choices listed. In some cases, more than one answer is correct.

1. Indicate which of the following molecules is chiral and how many stereocenters there are in each molecule.

a.

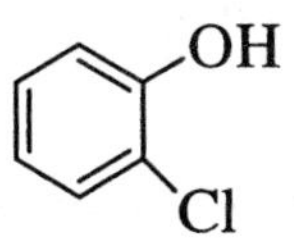

b.

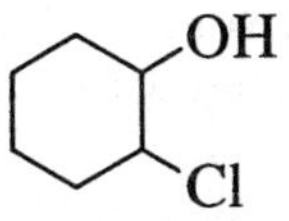

c.

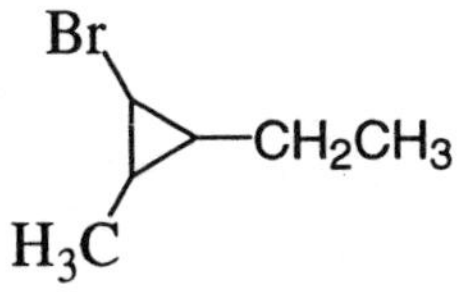

d.

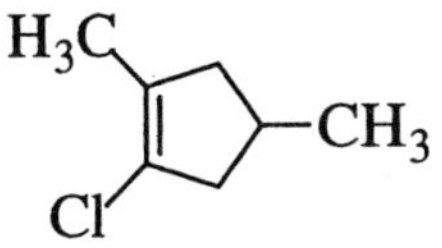

e.

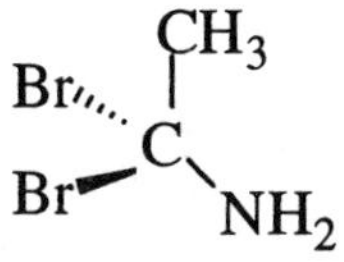

f.

CH3CHCHCH2CH3
| |
HO OH

2. Enantiomers have which of the following characteristics?

   a. rotate ordinary light
   b. have the same melting point
   c. have the identical molecular weight
   d. are superimposable mirror images

3. A racemic mixture exhibits which of the following characteristics?

   a. does not rotate plane polarized light
   b. rotates plane polarized light
   c. contains equal concentrations of both enantiomers
   d. often contains diastereomers

4. Indicate which of the molecules in question 1 will have an enantiomer.

5. Indicate which of the following molecules are enantiomers and which ones are diastereomers.

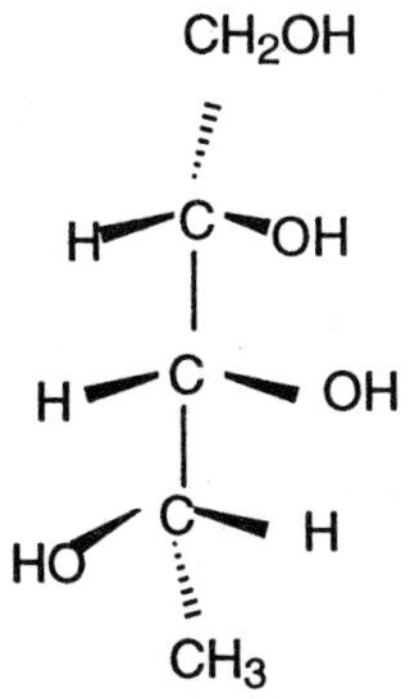

(i)

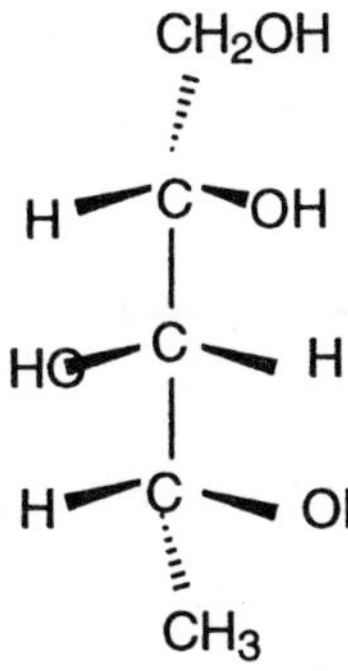

(ii)

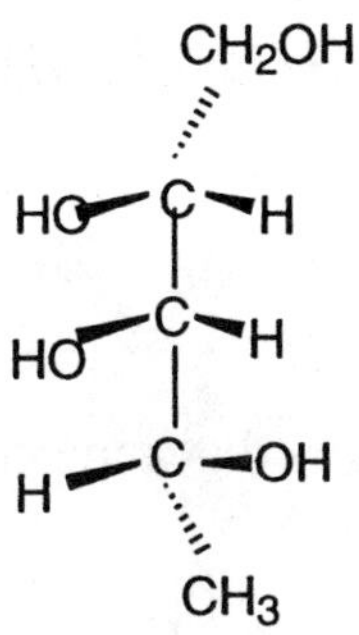

(iii)

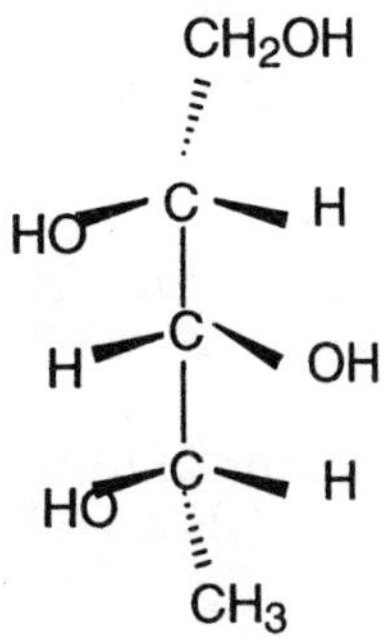

(iv)

6. The priority of the groups, CH_3, COOH, NH_2 and OH in the R, S system, is

   a. CH_3 > COOH > NH_2 > OH
   b. OH > NH_2 > COOH > CH_3
   c. COOH > CH_3 > NH_2 > OH
   d. COOH > OH > NH_2 > CH_3

7. Indicate whether the isomer shown is the R-form or the S-form.

a.

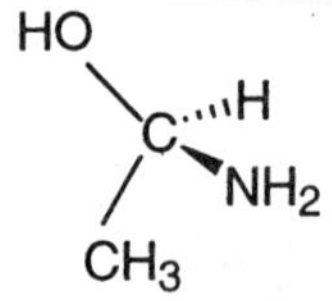

b.

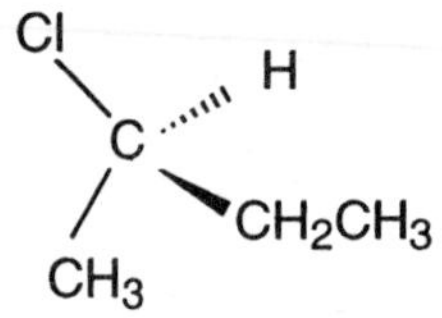

c.

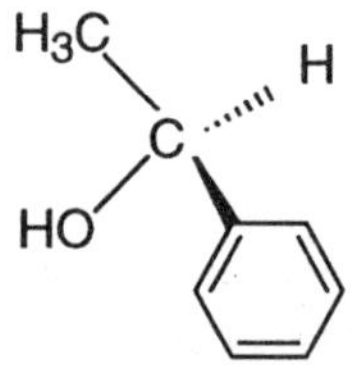

d.

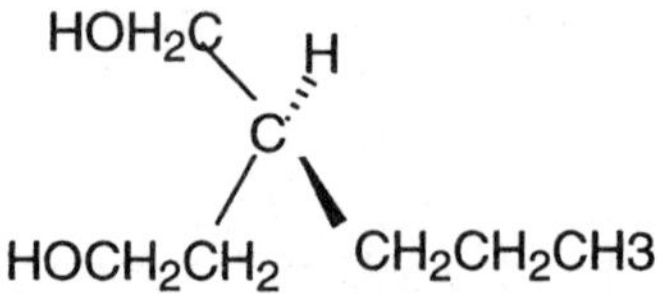

8. Calculate the maximum number of isomers in the molecules shown in questions 1 and 7.

<u>Question 1</u>

1a   1b
1c   1d
1e   1f

<u>Question 7</u>

7a   7b
7c   7d

<u>COMPLETION.</u> Complete the following statements with the appropriate word, phrase or number.

1. Enantiomers are ____________ mirror images. Because each enantiomer has at least one stereocenter, enantiomers must occur in ___________.

2 The maximum number of stereoisomers that can occur for the molecule, $CH_3CH(OH)CH(OH)CH(OH)CHO$, is ______.

3. Diastereomers are ___________________, but they are not ____ _____ of each other.

While the melting points for enantiomers are the same, the melting points for diastereomers are _______________.

4. Thalidomide is perhaps the most notorious drug in recent history and one that involves enantiomers. To reduce cost, the drug was sold as a racemic mixture of the enantiomeric pair. Although this antidepressant drug was not approved for use in the United States, it was widely used in Canada and Europe. One enantiomer was very effective as an antidepressant. However, pregnant women who took the drug were also subjected to the other enantiomer, which was mutagenic (causes changes in DNA) and antiabortive. As a result, many women who took the drug gave birth to severely deformed children. The enantiomeric pair is shown below.

Indicate how many stereocenters there are and identify the R- and the S-forms.

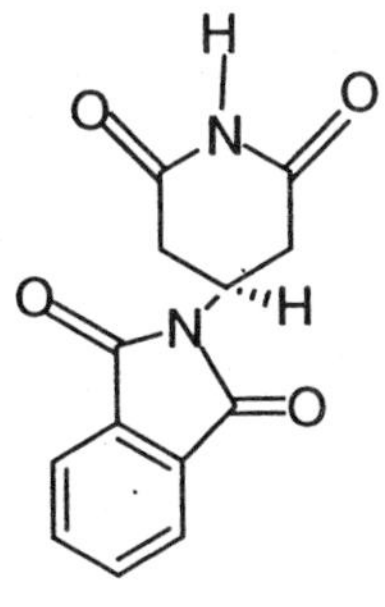

(i)

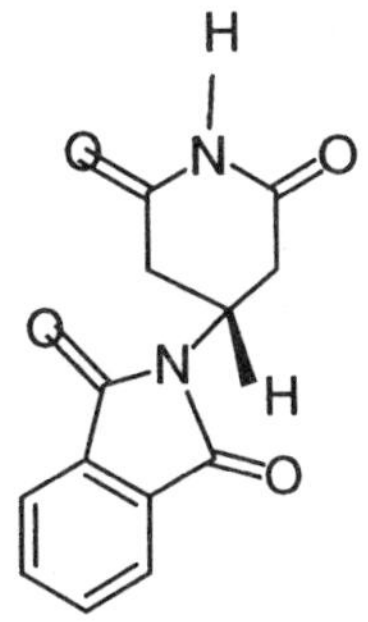

(ii)

5. Many biological molecules found in living systems are chiral. The steroid sex hormones are derived from cholesterol. Testosterone is responsible for the development of the secondary sex characteristics of males, while estrogens, such as estradiol, are required for the secondary sexual characteristics of females.

Indicate the number of stereocenters in testosterone and estradiol and the maximum number of stereoisomers possible for each.

Indicate which molecule contains an aromatic ring.

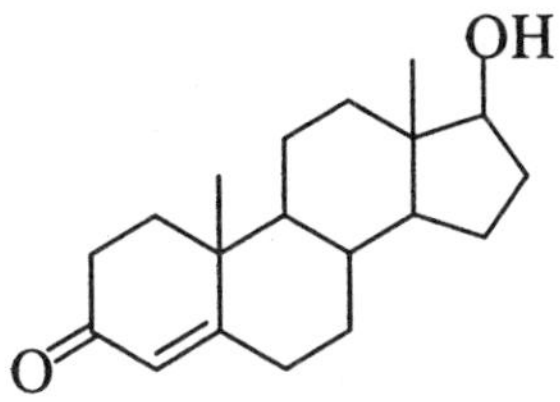

testosterone

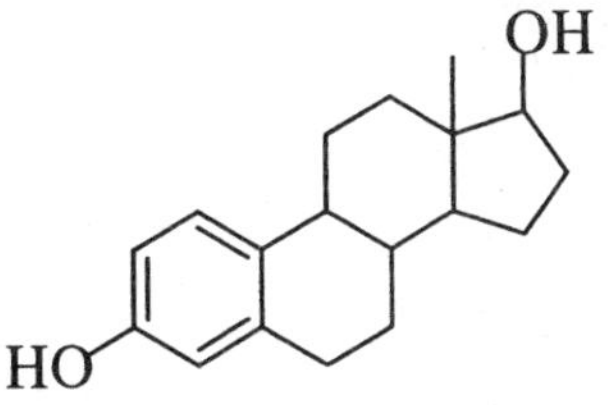

estradiol

ANSWERS TO SELF-TEST QUESTIONS

Multiple Choice

1.
   - a. achiral, 0
   - b. chiral, 2
   - c. chiral, 3
   - d. chiral, 1
   - e. achiral
   - f. chiral, 2
2. b, c
3. a, c
4. all chiral molecules - b, c, d & f
5. enantiomers; i & iii; ii & iv
   diastereomers; i ⅈ i & iv; ii & iii; iii & iv
6. b
7. a. R
   b. R
   c. S
   d. S
8. From question 1
   - a. 0
   - b 4
   - c. 8
   - d. 2
   - e 0
   - f. 4

   From question 7
   - a. 2
   - b. 2
   - c. 2
   - d. 2

Completion

1. nonsuperimposable, pairs
2. 8
3. stereoisomers, mirror images, different
4. There is one stereocenter in each enantiomer.
   (i) is R-thalidomide; (ii) is S-thalidomide
5. Testosterone has 6 stereocenters and a maximum of 64 possible stereoisomers
   Estradiol has 5 stereocenters and a maximum of 32 possible stereoisomers.
   Estradiol contains an aromatic ring.

"Common sense is the collection of
prejudices acquired by age eighteen"
- Albert Einstein

16

Amines

CHAPTER OBJECTIVES

After you have studied the chapter and worked the assigned exercises in the text and the study guide, you should be able to do the following.

1. Write out a general structural formula for a primary, secondary and tertiary amine and for a quaternary ammonium salt.

2. Write out the names for simple amines, including aliphatic, cyclic and aromatic compounds.

3. Rationalize the trend in the boiling points and water solubility for simple primary amines.

4. Describe the hydrogen bonding that takes place (a) between amines and (b) between amines and water and discuss the influence of hydrogen bonding on the physical properties of amines.

5. Give the names and structures for natural and synthetic amines that are used medicinally.

6. Write out the chemical equations for the reaction of amines with inorganic- and carboxylic acids.

7. Describe how the presence of an aromatic or aliphatic group in an amine influences its base strength.

8. Define the important terms and comparisons in this chapter and give specific examples where appropriate.

IMPORTANT TERMS AND COMPARISONS

Aromatic and Aliphatic Amines
Primary, Secondary and Tertiary Amines
Heterocyclic Amines, Heterocyclic Aliphatic Amines and Heterocyclic Aromatic Amines
Amphetamines (Pep Pills)
Alkaloids
Quaternary Ammonium Ion
Hydrogen Bonding Potential
N-Substituted Amines
Organic Bases
Kb and pKb
Acid/Base Chemistry of Amines
Benzodiazepine and Tranquilizers

SELF-TEST QUESTIONS

MULTIPLE CHOICE. Select the correct answer from the choices listed. In some cases, more than one choice will be correct.

1. The name of the compound given below is:

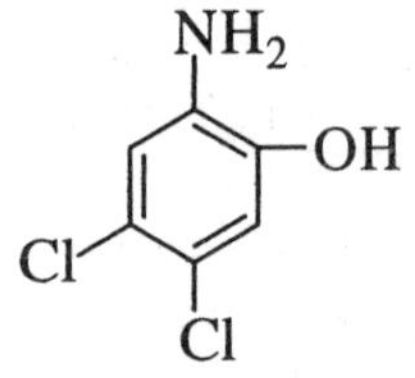

a. 2-Amino-4, 5-dichlorophenol
b. Dichlorohydroxylaniline
c. 2-Amino-4, 5-dichlorocyclohexanol
d. 3, 4-Dichloro-1-hydroxylaniline

2. The structure of N,N-Dimethylaniline is:

a.

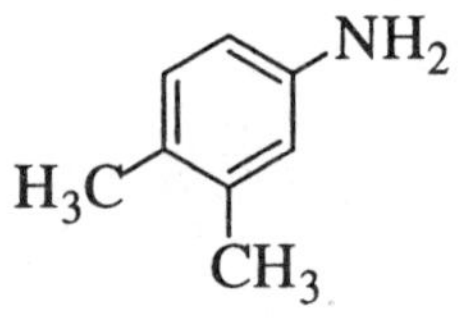

b.

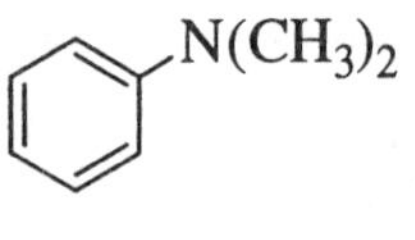

c.

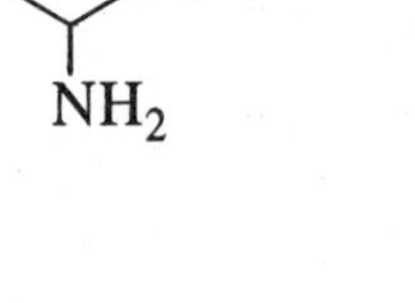

d.

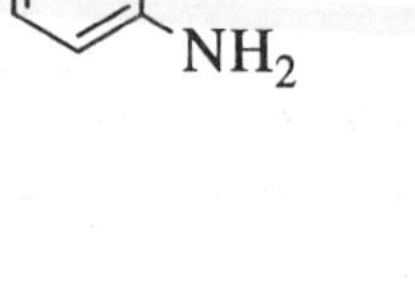

3. Protonation of amines that are used as drugs has the advantage that it

a. increases solubility
b. inhibits oxidation
c. decreases reduction
d. increases basic character

4. The pKb or Kb value for some basic amines are given below. Indicate which amine is the strongest base.

a. $pK_b = 4.5$
b. $K_b = 1 \times 10^{-8}$
c. $pK_b = 7.0$
d. $K_b = 2 \times 10^{-3}$

5. Which compound has the highest boiling point.

a. $CH_3CH_2CH_2OH$
b. $CH_3CH_2CH_2NH_2$
c. $CH_3CH_2OCH_3$
d. $CH_3NHCH_2CH_3$

6. Benzodiazepine, dopamine, epinephrine, procaine are all examples of:

a. chiral molecules
b. alkaloids
c. aromatic compounds
d. synthetic pain killers

7. A potential drug is made from a newly synthesized aromatic amine ($pK_b = 8.4$). Calculate the ratio of [amine/protonated amine] that will be present in the blood (pH = 7.4). You may use the Henderson-Hasselbalch equation.

a. 100/1
b. 90/1
c. 50/1
d. 10/1

<u>COMPLETION</u>. Answer or complete the questions as indicated.

1. Indicate which compounds is an aliphatic amine, aromatic amine, heterocyclic amine, aromatic heterocyclic amine or a combination of these, or none of these.
Indicate also whether it is a primary, secondary or tertiary amine.

a.

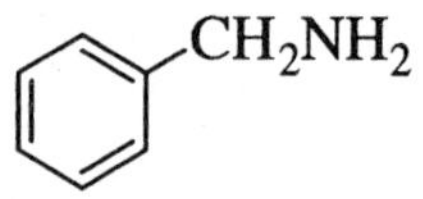

b.

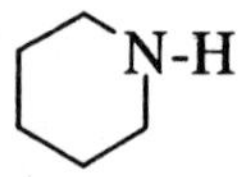

c.

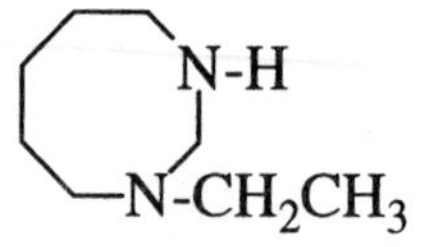

d.

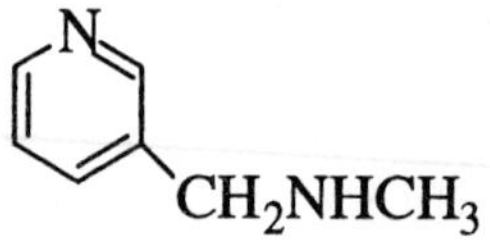

e.

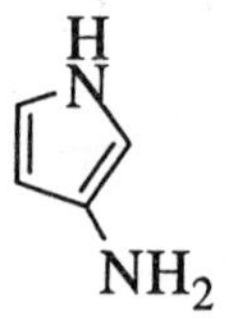

f.

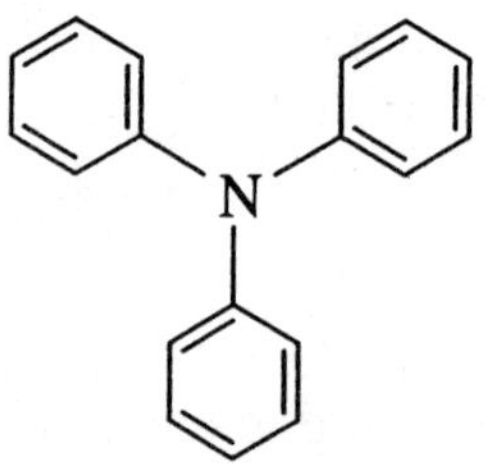

2. Draw these compounds.

   a. Ethylpropanamine
   b. Cyclohexanamine
   c. N-Methylaniline
   d. 1-Amine-4-chlorocyclohexane
   e. 5-Methylheptanamine
   f. Diphenylamine

3. Name these compounds

a.

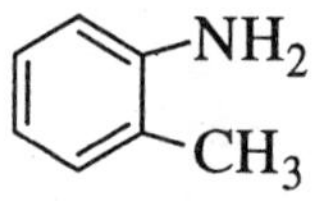

b.

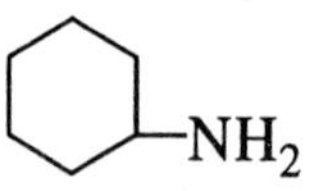

c.

H2NCH2CH2CH2CH2NH2

d

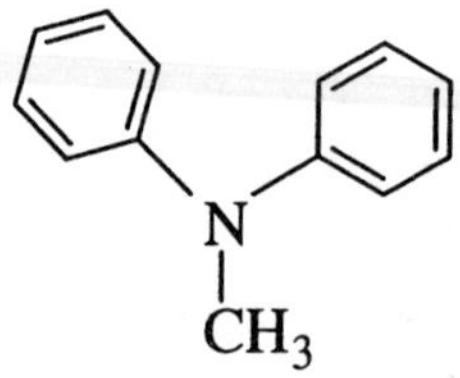

4. Indicate the order of increasing solubility of these compounds in water.

NH_3, NH_2CH_3, $N(CH_3)_2(C_6H_5)$

_______________ < _______________ < _______________

5. Complete the equations for the following acid/base reactions. If no reaction occurs, indicate no reaction.

a. $(CH_3)_3N$ + HCl ------->

b. $(CH_3CH_2)_2H_2^+$ + OH^- ------->

c. $N(CH_3)_4^+ Cl^-$ + NaOH ------->

6. a. Secondary amines have _______________ hydrogen(s) bound directly to the nitrogen.

b. The compound $[(CH_3CH_2)_3NH]^+Br^-$ is a _______________ and can be made by the reaction of _______________ with ____________________.

7.
<u>MATCHING</u>. Match the term in the right-hand column with the most appropriate answer in the left column.

1.	Weakest class of amine bases	a.	"Poison hemlock"
2.	Amphetamine	b.	Tranquilizer
3.	Pyridine	c.	Basic nitrogen-containing compounds extracted from plants
4.	Methylamine	d.	Pep pills, action similar to that of epinephrine
5.	Librium or Valium	e.	Water-soluble ammonium salt
6.	Methadone	f.	Aromatic amine
7.	Alkaloid	g.	Heterocyclic aromatic amine
8.	(S)-Coniine	h.	Aliphatic amine

ANSWERS TO SELF-TEST QUESTIONS

Multiple Choice

1. a
2. b
3. a, b
4. d
5. a
6. c
7. d

Completion

1.
   a. A primary, aliphatic amine (N is attached to only an aliphatic carbon)
   b. A secondary, aliphatic heterocyclic amine (contains no aromatic groups)
   c. Two aliphatic amines, one is a secondary amine, while the other is a tertiary amine
   d. A secondary, aliphatic amine and a heterocyclic, aromatic amine
   e. A primary, aliphatic amine and a secondary, heterocyclic aromatic amine
   f. a tertiary, aromatic amine

2.

a. $CH_3CH_2NHCH_2CH_2CH_3$

b.

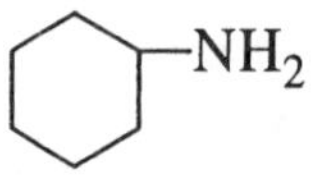

c.

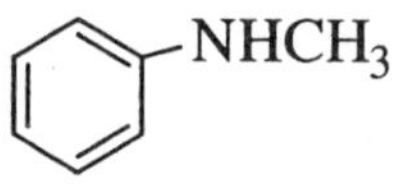

d.

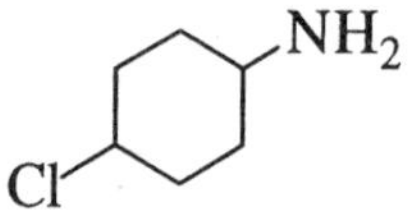

e.

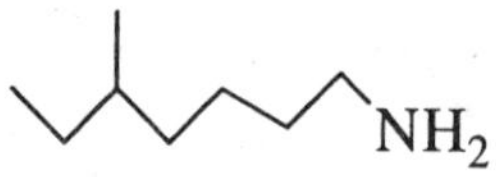

f.

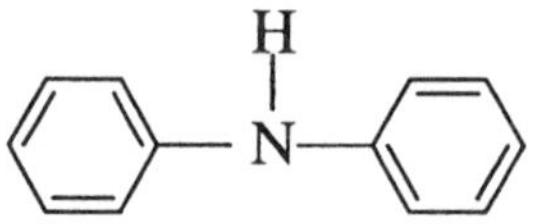

3.
   a. o-Methylaniline (o-Toluidine)
   b. Cyclohexanamine or cyclohexylamine
   c. 1, 4-Butandiamine
   d. Methyldiphenylamine

4. a. $N(CH_3)_2(C_6H_5) < NH_2CH_3 < NH_3$

5.

a. $(CH_3)_3N + HCl \text{ -------> } (CH_3)_3NH^+ Cl^-$

b. $(CH_3CH_2)_2H_2^+ + OH^- \text{ -------> } (CH_3CH_2)_2NH + H_2O$

c. $N(CH_3)_4^+ + NaOH \text{ -------> }$ No reaction

6. a. one
   b. quaternary amine or quaternary ammonium salt.
   Reaction of $(CH_3CH_2)_3N$ + HBr

Matching

1. f
2. d
3. g
4. h
5. b
6. e
7. c
8. a

Man's mind stretched by a new idea
never goes back to its original dimensions.
- paraphrase from Oliver Wendell Holmes

17

Aldehydes and Ketones

CHAPTER OBJECTIVES

After you have studied the chapter and worked the assigned exercises in the text and the study guide, you should be able to do the following.

1. Write out the characteristic functional group for an aldehyde and a ketone and contrast this to the structure and bond types found in functional groups presented in previous chapters.

2. Given the structural formula for an aldehyde or ketone, name the compound with the IUPAC nomenclature.

3. Given the IUPAC or common name for an aldehyde or ketone, write out its structural formula.

4. Choosing organic compounds of approximately the same molecular weight, compare the physical properties - boiling point and solubility in water - for an alkane, alcohol, amine, ketone, aldehyde, and ether.

5. Referring back to Chapter 14, write out reactions for the preparation of aldehydes and ketones that are derived from the oxidation of 2, 3 and 4 carbon alcohols.

6. Write out the balanced reactions for a positive Tollens' test (silver-mirror test).

7. Describe the basis for using the Tollin's test to distinguish between an aldehyde and ketone.

8. Clearly write out the general **structures** for a hemiacetal and an acetal that are derived from the reaction of an alcohol and an aldehyde. Provide at least one specific example of each.

9. Draw the general **structures** for a hemiacetal and an acetal that are derived from the reaction of an alcohol and a ketone. Provide at least one specific example of each.

10. Write the **chemical reactions** for the formation of a hemiacetal and an acetal that are derived from a reaction of an aldehyde. Provide examples using at least two different aldehydes.

11. Write the **chemical reactions** for the formation of a hemiacetal and an acetal that are derived from a reaction of a ketone. Provide examples using at least two different ketones.

12. Using a single molecule that contains both an alcohol and an aldehyde (ketone) functional group, show how a cyclic hemiacetal, and then on further reaction, a cyclic acetal, is formed.

13. Write out the equations for the reduction of aldehydes and ketones to produce a primary and secondary alcohol, respectively.

14. Summarize the oxidation, reduction and addition reactions for aldehydes and ketones.

15. Define the important terms and comparisons in this chapter and give specific examples where appropriate.

IMPORTANT TERMS AND COMPARISONS

Carbonyl Group
Saturated and Unsaturated Aldehydes
Sufixes -al and –one
Oxidation of Aldehydes
Tollens' Test - the Silver Mirror Test
Hemiacetal and Acetal (From a Ketone)
Keto and Enol Forms
Reducing Agents - $NaBH_4$ and NADH

Aldehyde and Ketone
IUPAC Nomenclature
Primary and Secondary Alcohols
Reduction of Aldehydes and Ketones
Hemiacetal and Acetal (From an Aldehyde)
Linear and Cyclic Acetals
Tautomerism and Tautomers
Hydride Anion (H^-)

In Chapters 10-17, we have increasingly expanded the coverage of organic chemistry - the study of carbon containing compounds with various functional groups (i. e., reactive groups), with an emphasis on the characteristic physical properties and chemical reactions associated with each group. The major functional groups presented to this point are the following.

MAJOR FUNCTIONAL GROUPS

I. Contains Only C and H
(Hydrocarbons)

A. Saturated

```
          |  |
Alkanes  -C-C-
          |  |
```

II. Contains C, H and O

A. Carbonyl Groups

```
             O
             ||
Aldehyde   R-CH

             O
             ||
Ketone     R-C-R'
```

B. Unsaturated

Alkene 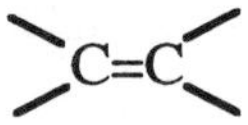

Alkyne $-C \equiv C-$

C. Aromatic

B. Others

Alcohol R-OH

Ether R-O-R'

Phenol 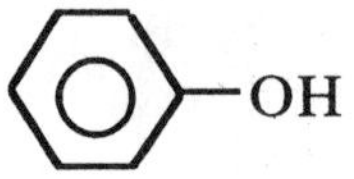

III. <u>Contains C, H and S or N</u>

Thiol or Mercaptan R-SH

Disulfide R-S-S-R'

Amines NH_nR_{3-n}

Organize and review the summary sheets of the reactions that characterize each functional group. The importance of making and continually reviewing (even casually) these sheets throughout the course will become vividly clear when you carry out a major review of the material for each hourly exam and the final exam.

This chapter focuses on aldehydes and ketones, their physical properties and the chemical reactions associated with each functional group. The production of hemiacetals and acetals from the reaction of an aldehyde or a ketone with an alcohol is of special importance here and in the subsequent chapter on carbohydrate chemistry. The following table summarizes these general structures. You should be able to readily recognize and distinguish between these six compounds before going onto the SELF-TEST.

<u>Aldehyde</u>		**<u>Hemiacetal</u>**	**<u>Acetal</u>**
O ‖ **R-C-H**	⟶	OH | **R-C-H** | OR'	OR' | **R-C-H** | OR'
<u>Ketone</u>		<u>Hemiacetal</u>	<u>Acetal</u>
O ‖ **R-C-R"**	⟶	OH | **R-C-R"** | OR'	OR' | **R-C-R"** | OR'

Consider the reaction of an adehyde or a ketone with an alcohol to produce the corresponding hemiacetal or acetal shown in the table above. Note that the **<u>bold-type groups</u>** remain in the structure for the three related compounds (reactant and either product). The R primed group (R') comes from the alcohol group that reacts with the aldehyde or ketone. Write out the reactions for different alcohols, reacting with an aldehyde or ketone, showing the structure of the product. Indicate from which reactant each of the R groups is derived.

SELF-TEST QUESTIONS

MULTIPLE CHOICE. Select the correct answer or answers from the choices listed. In some cases, more than one answer is correct.

1. Mild oxidation of an aldehyde produces a carboxylic acid. However, mild oxidation of a ketone results in:

   a. an ether
   b. secondary alcohol
   c. carboxylic acid
   d. no reaction

2. Identify the compounds as either an aldehydes, hemiacetals or acetals.

   a.

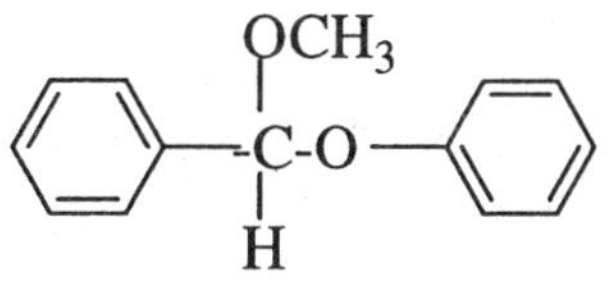

   b.

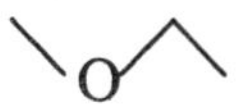

   c.

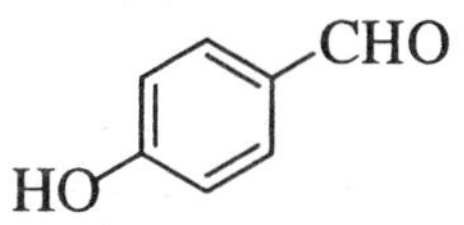

   d.

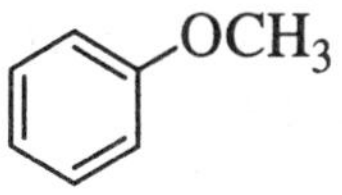

   e.

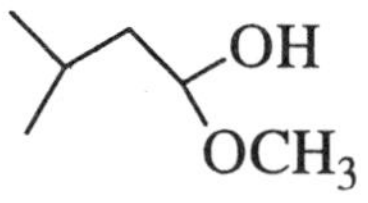

   f.

   O
   ||
   CH3CH2-C-CH2CH3

3. Select compounds in question 2 that have a carbonyl group.

4. The reaction of an alcohol with a ketone is an example of a(n):

   a. elimination reaction
   b. addition reaction
   c. hydrolysis reaction
   d. substitution reaction

5. Ketones are prepared by the oxidation of:

   a. secondary alcohol
   b. primary alcohol
   c. tertiary alcohol
   d. all of these

6. The characteristics that are produced in a positive Tollens' test are:

   a. Oxidation of an aldehyde and reduction of $Ag(NH_3)_2^+$.
   b. Reduction of an aldehyde and oxidation of $Ag(NH_3)_2^+$.
   c. Oxidation of a ketone and production of a silver mirror.
   d. All of the above.

7. The structural formula for benzaldehyde is:

a.

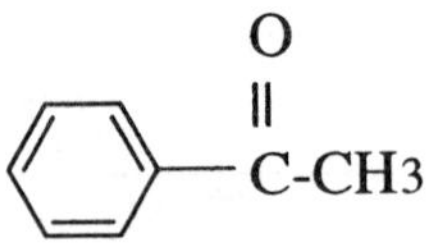

b.

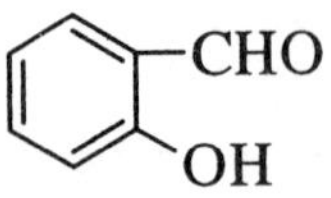

c.

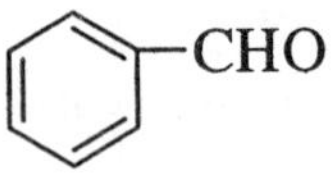

d.

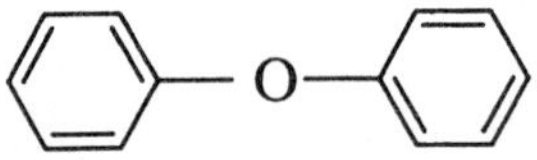

8. Indicate which of the following molecules can undergo a keto-enol equilibrium.

a. CH_3CH_2-OH

b. $CH_3CH_2CH_2COCH_3$

c.

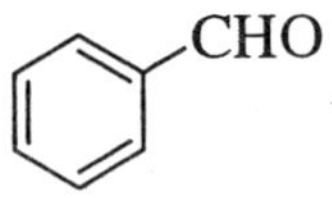

d.

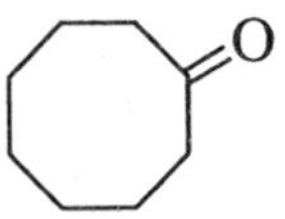

9. The structural formula for 2,3-Dimethylcyclohexanone is:

a.

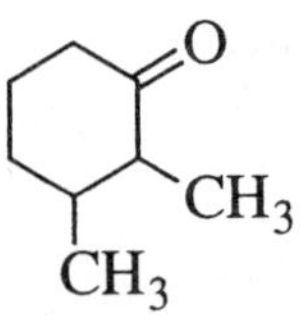

b.

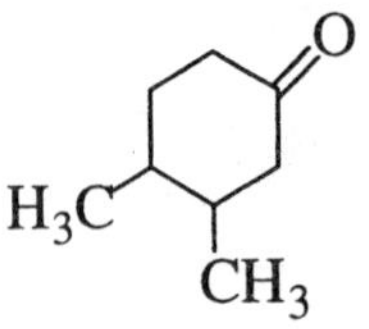

c.

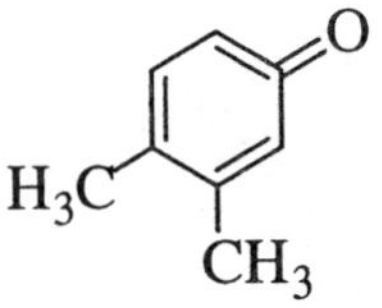

d.

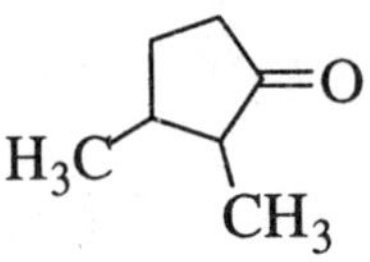

H_3C Br
| |
10. The IUPAC name for $H_3CCCHCH_2CH_2CH_2CHO$
|
Br

a. 5, 6-Dibromo-6-methylheptanal
b. 1, 2-Dibromo-1-methylheptanone
c. 5, 6-Dibromo-1-methylheptanal
d. 1 ,2-Dibromo-1-methylheptanal

11. The structure for one of the forms of the nucleic acid base, guanine, is shown below. This form is:

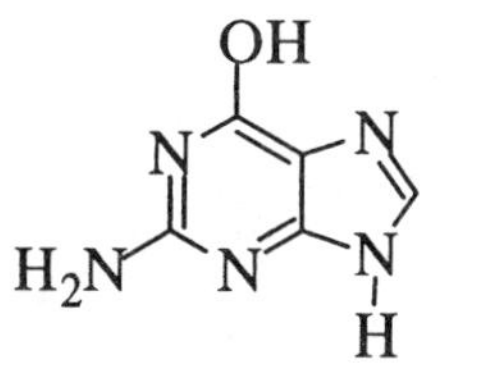

a. a ketone
b. the keto form of guanine
c. the enol form of guanine
d. an acetal

12. Indicate whether the product of an oxidation or a reduction (indicate) of the molecules below will be an aldehyde or a ketone.

a.

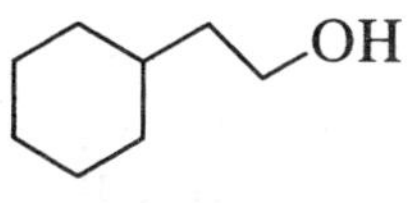

c.

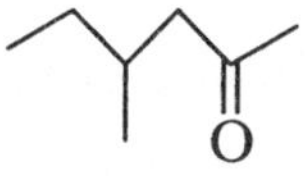

b.

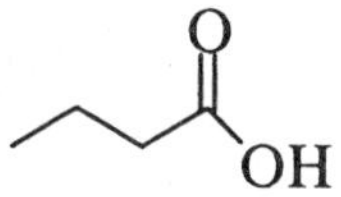

d.

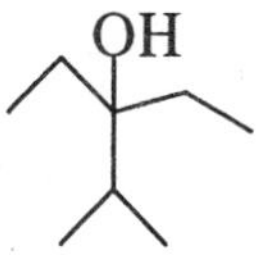

COMPLETION. Write the word, phrase or number in the blank or draw the appropriate structure in answering the question.

1. Write out the IUPAC names for the following molecules.

a.

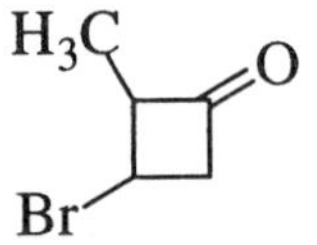

b.

CH3
|
CH3-C-CHO
|
CH3

c.

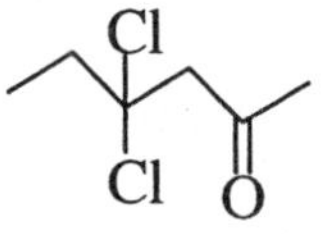

d.

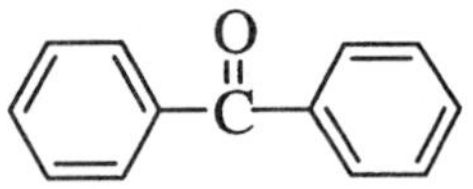

e.

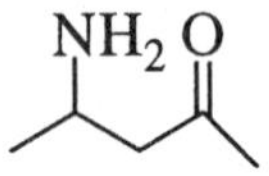

2. Given compounds of comparable molecular weight, the order of increasing boiling points for an alcohol, alkane, ether and aldehyde is:

_______________ > _______________ > _______________ > _______________

3. The common name for one of the simplest aldehydes, CH3CH, is _______________.
||
O

4. The general structure of an aldehyde would suggest that this class of compounds ____________ hydrogen bond to each other.

5. The hydrolysis of an acetal yields an _______________ and _______________.

O
||
6.. The reaction of a ketone, R-C-R", with two molecules of an alcohol, R'-OH, yields an ____________, which contains two ether linkages and the general structure is

____

7. Complete the following equations. [H] and [O] indicate reduction and oxidation, respectively. If no reaction occurs, indicate just that.

a.

b.

c.

d.

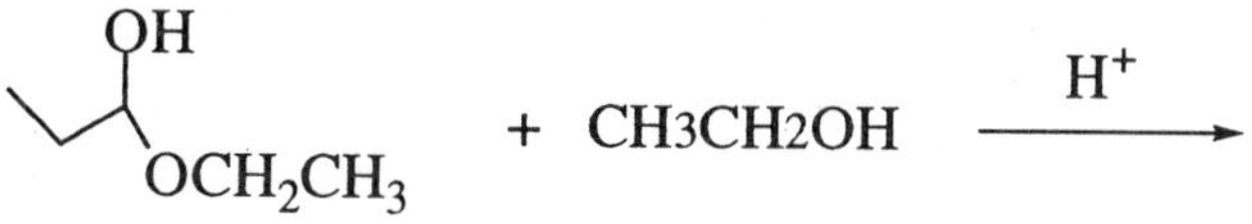

e.

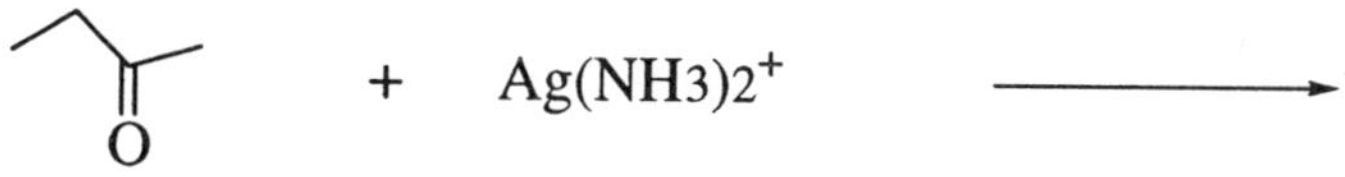

f.

8. The oxidation of (C_6H_5)-CH_2CHO yields a molecule with the structural formula

____

9. In an reduction reaction of an aldehyde with $NaBH_4$, a __________ anion, which is a hydrogen with _____ electrons, attacks the carbonyl ______ of the aldehyde functional group.

10. The mixing of an aldehyde with an acid or base yields ______________ reaction because all aldehydes are ______________ acidic or basic in character.

OH
|
11. Draw the (a) keto form of the compound, $H_3CC{=}CHCH_3$ and (b) the enol form for the compound

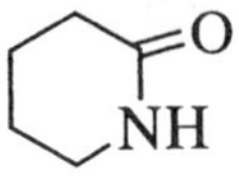

ANSWERS TO SELF-TEST QUESTIONS

<u>Multiple Choice</u>

1. d
2. a (acetal), c (aldehyde), e (hemiacetal)
3. c, f
4. b
5. a
6. a
7. c
8. b, d
9. a
10. a
11. c
12. a (oxidation to an aldehyde)
    b (reduction to an aldehyde)

<u>Completion</u>

1. a. 3-Bromo-2-methylcyclobutan-1-one
   b. 2, 2-Dimethylpropanal
   c. 4, 4, Dichlorohexan-2-one
   d. Diphenylketone
   e. 4-Amine-2-pentanone
   (Answers a, c and e are alternate, acceptable forms to naming ketones)

2. alcohol > aldehyde > ether > alkane
3. acetaldehyde
4. does not
5. alcohol and aldehyde (or ketone)

OR'
|
6. acetal, R-C-R", in which the R' groups are derived from the alcohol
|
OR'

7.

a.

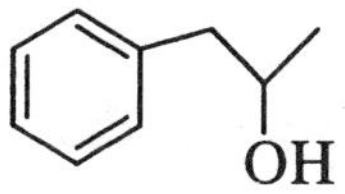

b.

 or 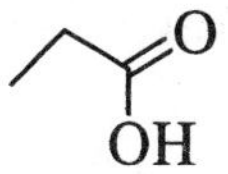

c.

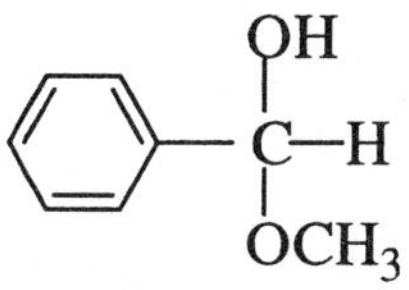

d.

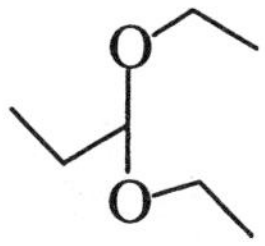

e. no reaction

f.

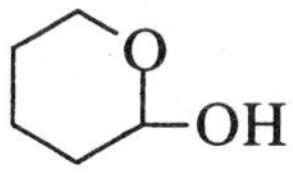

8.

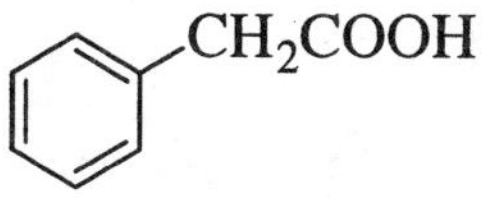

9. hydride, 2, carbon

10. no, neither

11.

a.
```
        O
        ||
CH3CCH2CH3
```

b.

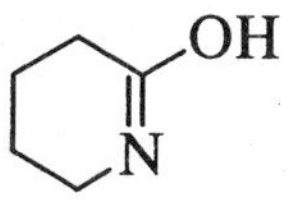

"ASK WHY"

- Sign in the office of Kenneth W. Kirk,
former Associate Commissioner, FDA

18

Carboxylic Acids, Anhydrides, Esters and Amides

CHAPTER OBJECTIVES

After you have studied the chapter and worked the assigned exercises in the text and the study guide, you should be able to do the following.

1. Recognize the characteristic functional groups in carboxylic acids, the anhydrides derived from carboxylic acids and phosphoric acid, carboxylic esters, phosphate esters and amides.

2. Write out the structures for these compounds, in addition to those for a carboxylate salt, including all atoms in the structures.

3. Draw the structure of the carboxylic acids containing one to six carbons, write out their common names and indicate three characteristics of carboxylic acids.

4. Explain the influence of hydrogen bonding in the physical characteristics of carboxylic acids.

5. Write the reaction for the dissociation of a carboxylic acid and also for the reaction with a strong base, such as NaOH.

6. Given the common or IUPAC name for a carboxylic acid, ester or amide, write out its structural formula.

7. Given the condensed structural formula for a carboxylic acid, ester or amide, name the compound by either its common or IUPAC name.

8. Contrast the physical properties of carboxylic acids with the comparable esters and amides.

9. Write out the complete equation for the reaction of a carboxylic acid with:
   a. a base to form a salt of a carboxylic acid
   b. an alcohol to form an ester
   c. another carboxylic acid to form an (carboxylic) anhydride

10. Contrast the reactions and products associated with the acid or base catalyzed hydrolysis of esters.

11. Write out structures for the anhydrides and esters of phosphoric acid and note the similarities to carboxylic anhydrides and esters.

12. Continue to write out summary sheets for the reactions associated with the new functional groups presented in this chapter. These should include:
   a. Carboxylic acids and the corresponding carboxylic esters, cyclic esters and anhydrides and the corresponding amides..
   b. Phosphoric acid and the corresponding phosphate esters and anhydrides.

13. Write out the reactions for the step-growth polymerization for polyamides, polyesters and polycarbonates.

14. Define the important terms and comparisons in this chapter and give specific examples where appropriate.

IMPORTANT TERMS AND COMPARISONS

Carbonyl Group and Carboxyl Group
Amines and Amides
Suffixes -oic, -dioic, -oate and -olactone
Hydrogen Bonding
Esters and Lactones
Anhydrides and Acyl-Groups
Esters and Lactones (Cyclic Esters)
Penicillins and β-Lactam Ring
Saponification
Aspirin (Acetylsalicylic Acid)
Barbiturates
Condensation and Step-Growth Polymers
Nylon 66, Kevlar, Dacron and Lexan

Carboxylic Acids, Esters and Anhydrides
Ka and pKa
Carboxylic Acid Dimer
Hydrophobic and Hydrophilic
Acid/Base Reactions
Fischer Esterification
Amides and Lactams (Cyclic Amides)
Ester and Amide Hydrolysis Reactions
Le Chatelier's Principle
Cyclooxygenase (COX) Enzymes
Phosphoric Anhydrides and Esters
Polyamides, Polyesters and Polycarbonates
Biodegradable

FOCUSED REVIEW

The two basic functional groups in this chapter, from which all other derivatives (compounds) are derived are:

1. Carboxylic acid

```
  O
  ||
R-C-OH
```

2. Phosphoric acid

```
   O
   ||
HO-P-OH
   |
   OH
```

The derivatives, together with the so-called "parent" compound from which they are derived, are shown below.

```
                    Carboxylic Acid                Phosphoric Acid
                          O                              O
                          ||                             ||
Acid                   R-C-OH                         HO-P-OH
                                                         |
                                                         OH

                          O            O           O  2-         O  3-
                          ||           ||          ||            ||
Anion(s)               R-C-O⁻       HO-P-O⁻      O-P-O         O-P-O
                                       |           |             |
                                       OH          OH            O

                          O            O           O             O
                          ||           ||          ||            ||
Ester(s)               R-C-OR'      HO-P-OR'   R'O-P-OR'     R'O-P-OR'
                                       |           |             |
                                       OH          OH            OR'

                                   monoester     diester       triester

                    O   O                          O   O
                    ||  ||                         ||  ||
Anhydride          R-C-O-C-R                    HO-P-O-P-OH
                                                   |   |
                                                   OH  OH

                                                  O   O
Ester of the                                      ||  ||
Anhydride            ______                     ⁻O-P-O-P-OR
                                                  |   |
                                                  O₋  O₋

                                               O   O   O
                                               ||  ||  ||
                                             ⁻O-P-O-P-O-P-OR
                                               |   |   |
                                               O₋  O₋  O₋
```

These general examples of phosphate esters will become familiar in subsequent chapters, in the form of specific molecules that serve as metabolic intermediates (phosphoenolpyruvate, glucose phosphate and others) and in ATP (adenosine triphosphate) hydrolysis for energy production.

```
                O
                ||
Amides       R-C-NHR                  ____________
```

In each of the derivatives, note exactly how the structure has been changed in going from the parent acid to the derivative compound. The similarities between the carboxylic acid and phosphoric acid should be clearly evident, along with the similarities in the analogous compounds derived from each parent. To become more familiar with these compounds and their chemistries, write out the complete equations for forming each derivative compound from the "parent" acid. Add these reactions to your summary sheets.

SELF-TEST QUESTIONS

MULTIPLE CHOICE. In the following exercises, select the correct answer from the choices listed.

1. The saponification of the carboxylic ester shown below yields the following products:

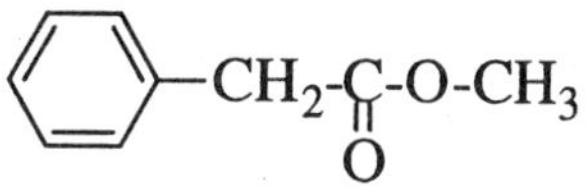

a.

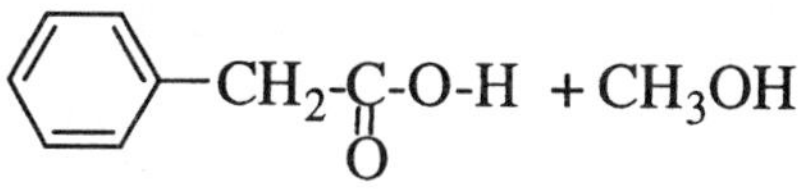

b.

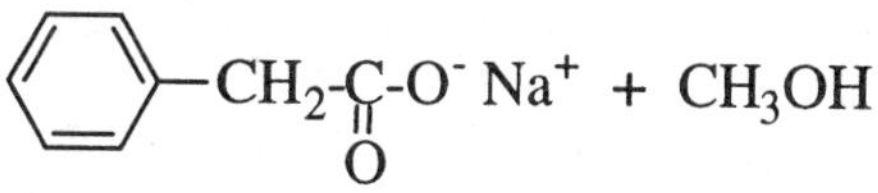

c.

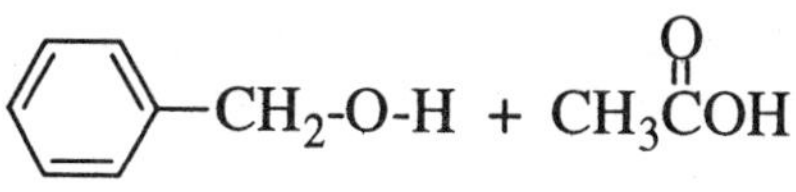

d. None of these

2. Carboxylate ion can exist in water at only:

   a. pH 7
   b. low pH
   c. high pH
   d. $[H^+]$ greater than 10^{-2} M

3. The name of the compound, $H_3CCH_2CH_2CH_2\overset{\overset{O}{\|}}{C}OCH_2CH(CH_3)_2$, is

   a. Isobutyl pentanoate
   b. Butyl propanoate
   c. Valeryl isobutyrate
   d. Isobutyl butanoate

4. Three organic acids are listed below with their pKa values.

		pKa value
Benzoic acid	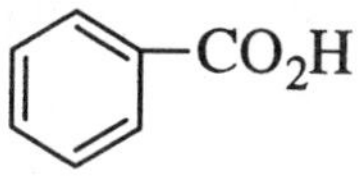	4.19
Chloroacetic acid	$ClCH_2COOH$	2.86

Formic acid	HCOOH	3.74

The order of decreasing acid strength is:

a. chloroacetic acid > formic acid > benzoic acid
b. benzoic acid > formic acid > chloroacetic acid
c. all are the same
d. not determinable from available data

5. The name of the compound,

```
                  O
                  ||
H3CCH2CHCH2CHC-OH,
        |    |
       CH3   Br
```

is

a. 1-Bromo-3-methylhexanoic acid
b. 2-Bromo-4-methylhexanoic acid
c. 2-Bromo-4-methylpentanoic acid
d. 5-Bromo-3-methylpentanoic acid

6. The hydrolysis of what ester produces butanoic acid and phenol.

a. Phenyl butanoate
b. Benzyl propanoate
c. Butyl benzoate
d. Phenyl propanoate

7. 5-Pentanolactam is an example of

a. an ester
b. an amide
c. a cyclic ester
d. a cyclic amide

8. The ethyl group in N-ethylbutanamide is bonded directly on the

a. carbonyl carbon
b. carbon 3
c. carbon 2
d. the amide nitrogen

COMPLETION. Write the word, phrase or number in the blank or draw the appropriate structure in answering the questions.

1. Write out the functional group or groups contained in each of the following condensed structural formulas. Some of the functional groups were presented in previous chapters.

```
   H3C
    |
a. H2CCH=CHCO2H
```

____

O OH
|| |
b. $H_3CC\text{-}CHCH_2C(CH_3)_3$ ____________________

c.

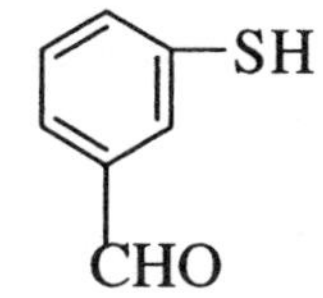

d.

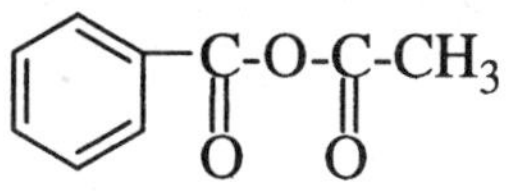

O
||
e. $(HO)_2POCH_2CH_3$ ____________________

O
||
f. $(H_3C)_2CHOCCH_3$ ____________________

g.

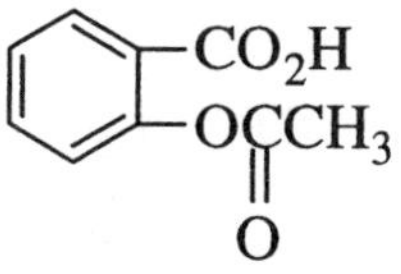

h. $H_3CCO_2^-$ ____________________

2. Complete the following equations and name the type of reaction.

a. ____________________ + NaOH -----> $H_3CCO_2^-$ Na^+ + H_2O

____

____

b.

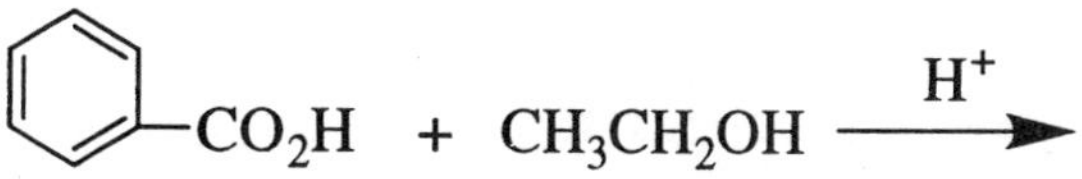

c. ____________________ $\xrightarrow{[O]}$ $(H_3C)_2CHCH_2CO_2H$

d. $H_3CCH_2CO_2H$ + __________ -----> $H_3C\overset{\overset{O}{\|}}{C}\text{-}O\text{-}\overset{\overset{O}{\|}}{C}CH_2CH_3$

e.

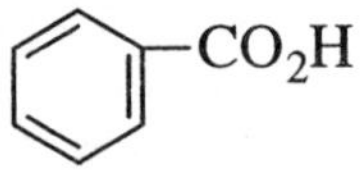

$\rightleftharpoons$ H^+ +

f.

__________ + __________ $\xrightarrow{H_2O}$ $CH_3CH_2CO_2H$ + NH_4Cl

3. The structure of the dimer formed by hydrogen bonding between 2 carboxylic acid molecules, RCO_2H, is:

4. Polyesters are examples of condensation polymers. To prepare these large molecules, each reactant molecule must have at least two functional groups. Draw out at least one repeating unit produced in the polyester formed in the reaction shown.

$HOCH_2OH + HO_2CCH_2CH_2CH_2CO_2H$ ----->

5. The order of decreasing boiling points for an alkane, a carboxylic acid, an ester and an alcohol, all of comparable molecular weight, is:

________ > ________ > ________ > ________

6. The primary reason for the order observed in the compounds in question 5 is ________________________________.

7. The molecular formula for n-octyl acetate is $H_3C\overset{\overset{\displaystyle O}{\|}}{C}O(CH_2)_7CH_3$. This molecule gives the fragrance of an orange. Draw the structure of the alcohol and the acid from which the ester was made.

________________ alcohol

________________ acid

8. Name the following compounds.

a.

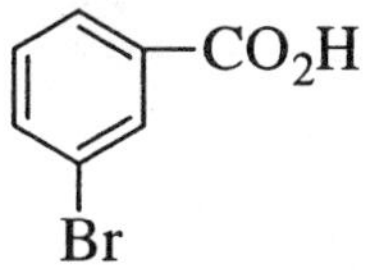

b.

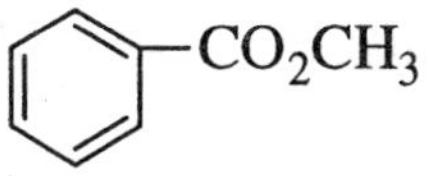

c.

Cl

|

$H_3CCH_2C\text{-}CO_2^-\ K^+$ ____________________

|

Cl

9. Draw the structures for the following molecules.

   a. 2-Phenylpropanoic acid
   b. Phenyl 2-methyl propionate or phenyl isobutyrate
   c. Isopropy benzoate
   d. 6-Hexanolactone
   e. N-Methyl-propanamide
   f. Ammonium pentanonate
   g. Methyl 3-phenylpentanonate

10. Aspirin is now known to __________ the synthesis of ______________ by reacting with the enzyme, ________________, which is often referred to by the acronym, COX.

ANSWERS TO SELF-TEST QUESTIONS

Multiple Choice

1. b
2. c
3. a
4. a
5. b
6. a
7. d
8. d

Completion

1. a. alkene, carboxylic acid
   b. ketone, alcohol (note that this is not a carboxylic acid)
   c. thiol, aldehyde, aromatic
   d. anhydride, aromatic
   e. phosphate (mono) ester
   f. (carboxylic) ester
   g. carboxylic acid, carboxylic ester
   h. carboxylate anion

2. a. H_3CCO_2H; neutralization of (a carboxylic acid)

   b.

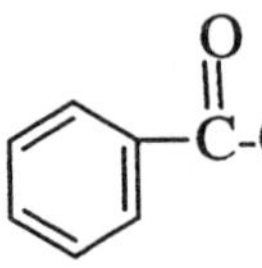

$\text{-C-OCH}_2\text{CH}_3$; Preparation of a (carboxylic) ester

c. $(H_3C)_2CHCH_2CH_2OH$; preparation of a carboxylic acid by oxidation of a primary alcohol or the corresponding aldehyde, $(H_3C)_2CHCH_2CHO$.

d. H_3CCO_2H; preparation of an anhydride

e. Reversible dissociation of a carboxylic acid (a weak acid)

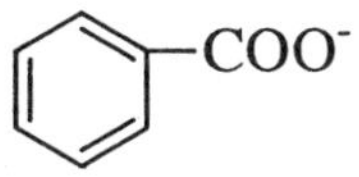


```
              O
              ||
f. CH3CH2C-NH2,  HCl     Acid hydrolysis of an amide
```

```
3.      O......HO-C-R
        ||        ||
     R-C-OH.......O
```

4. The following partial structure for the polymer shows one of the individual repeat units from the diol and the diacid underlined.

```
   O              O              O
   ||             ||             ||
-CC[O-CH2-O-CCH2CH2CH2C]-OCH2O-
     _______  ____________
```

5. carboxylic acid > alcohol > ester > alkane

6. hydrogen bonding between molecules. The greater the hydrogen bonding between molecules, the more energy it requires to vaporize the liquid and therefore the higher is the boiling point.

7. $H_3C(CH_2)_7OH$; H_3CCO_2H

8. a. 3-Bromobenzoic acid
   b. Phenyl acetate
   c. Potassium 2,2-dichlorobutanoate

9.

a.

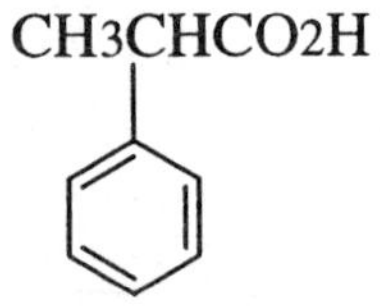

b.

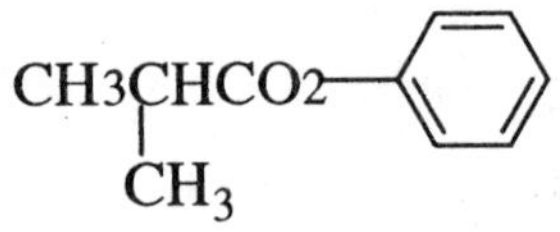

c.

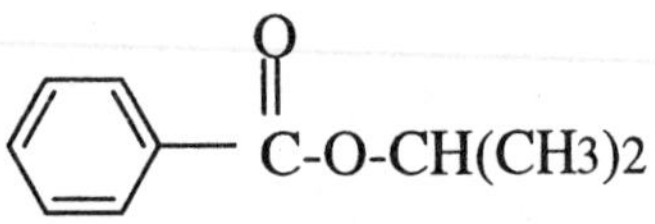

d.

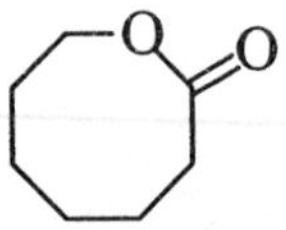

e.

$CH_3CH_2\text{-}\overset{O}{\overset{\|}{C}}\text{-}NHCH_3$

f.

$CH_3CH_2CH_2CH_2CO_2^- \, NH_4^+$

g.

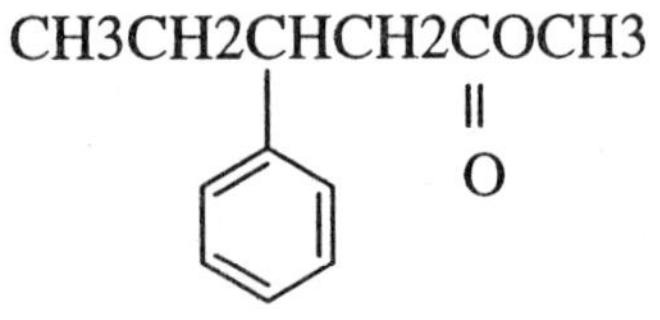

10. inhibit, prostaglandins, cyclooxygenase

Burnt Sugar Icing Recipe
2/3 cup granulated sugar
1/3 cup boiling water
3 cups powdered sugar
2 tablespoons butter
- H. Morgan
You Can't Eat That!

19

Carbohydrates

CHAPTER OBJECTIVES

After you have studied the chapter and worked the assigned exercises in the text and the study guide, you should be able to do the following.

1. Define carbohydrates chemically and indicate a number of specific uses of carbohydrates in bacteria, plants and humans.

2. Draw the D-form of the aldoses and ketoses that have three, four and five carbon atoms. Identify the stereocenter and the position that defines the sugar as being either the D- or the L-form.

3. Calculate the maximum number of stereoisomers (2^n rule) for aldoses and ketoses with 3, 4 and 5 carbon atoms. Draw the Fischer projection formulas for these and identify the enantiomeric pair of molecules and the diastereomers. Specify the D- and L-form of these sugars.

4. Indicate the physical and chemical properties that differ in enantiomers.

5. Describe in words the solution equilibrium for a monosaccharide such as β-D-glucose. Also draw the open chair and ring structures for the monosaccharides in the equilibrium. Identify the hemiacetal units in the Haworth formulas.

6. Write out the reactions for D-glucose undergoing (a) reduction, (b) oxidation and (c) phosphorylation. Note the atomic position at which the reaction takes place.

7. List three monosaccharides and three disaccharides of importance. Name the products obtained from the hydrolysis of the specific disaccharides.

8. Draw the Haworth structures for the three common disaccharides.

9. Draw the Haworth structure for maltose, pointing out (a) the acetal unit, (b) the reducing ring and (c) the glycosidic linkage.

10. Describe the structural difference between amylopectin and amylose. Draw a minimum unit of this structure to show the difference.

11. Describe the major structural difference between starch and cellulose. Explain the significance of this with regard to human metabolism.

12. Characterize the linkage types found in the hyaluronic acid and heparin. Explain why both polysaccharides are considered acidic.

13. Define the important terms and comparisons in this chapter and give specific examples where appropriate.

IMPORTANT TERMS AND COMPARISONS

Carbohydrates
Aldoses and Ketoses
Chiral Molecule
Penultimate Carbon
Anomeric Carbon Atoms and Anomers
Hemiacetals and Acetals
α- and β-D-Glucose
Alditols and Sorbitol
Reducing and Non-Reducing Sugar
Ascorbic Acid (Vitamin C)
Enantiomers and Diastereomers
Glucose, Dextrose, or Blood Sugar
Sucrose, Lactose and Maltose
Mutarotation
Saccharin and Artificial Sweetening Agents
Antigen and Antibody
Starch and Cellulose
α(1--->4) and β(1--->6) Glycosidic Linkages
Diabetes Mellitus
Acidic Polysaccharide
Hyaluronic Acid and Heparin
D- and L-Monosaccharides
Stereocenter
Fisher Projection
Galactosemia
α- and β-Configuration of the Anomeric Carbon
Glycoside and Glycosidic Bond
Aldonic Acids
Chair Conformation for Cyclic Sugars
Uronic Acids
Glycoside and Glucoside
Disaccharides and Polysaccharides
"Locked" Ring of a Disaccharide
Pyranose and Furanose Rings
A, B, AB and O Blood Groups
Oligo- and Polysaccharides
Amylose, Amylopectin and Glycogen
Hydrolysis and Condensation Reactions
Cellulose and β-Glucosidases
Glucose Oxidase
Enediol Intermediate

FOCUSED REVIEW

This chapter begins the formal coverage of **biochemistry** (concepts), including the nomenclature, structures, characteristics and reactions associated with these vital molecules. As will become evident, virtually all of these molecules, whether essential small molecules or biomacromolecules, have one, and in most cases, many stereocenters. As you study this and subsequent chapters, it is important to review Chapter 15 (Chirality:The Handedness of Molecules) and realize that this is an essential property of most molecules in living systems.

In addition, Chapter 17 (Aldehydes and Ketones) provides background for carbohydrates since, by definition, carbohydrates are polyhydroxyaldehydes, polyhydroxyketones or compounds that yield them after reaction with water (hydrolysis). Review these previous chapters as this new material is being assimilated.

Structures and Reactions

The cyclic forms of most monosaccharides contain a **hemiacetal** unit formed by the internal reaction of the alcohol group reacting with the carbon atom of the aldehyde group. Disaccharides contain an **acetal** unit and, if it includes a reducing sugar unit, it will also contain a hemiacetal unit. The formation and stability of these saccharide units is a major focus of this chapter. As introduced in Chapter 17, the reaction of an aldehyde or a ketone with an alcohol yields the formation of a hemiacetal.

a.

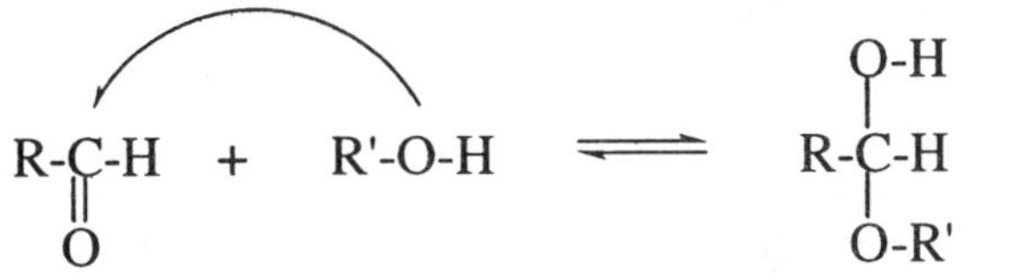

aldehyde alcohol hemiacetal

b.

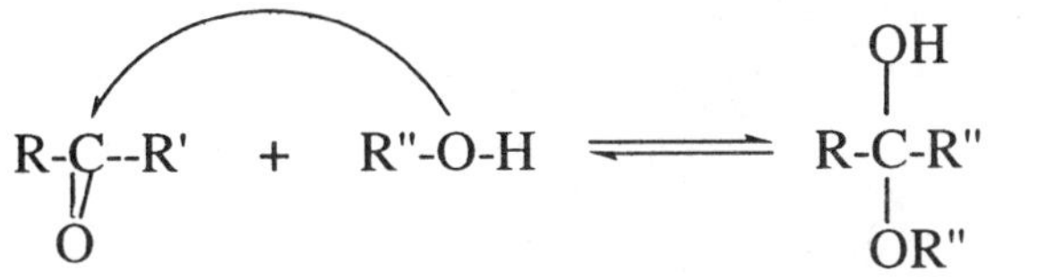

ketone alcohol hemiacetal

Recall that hemiacetals are unstable unless cyclic systems are formed as in the 5- or 6-membered furanose or pyranose rings, respectively.

For D-glucose, note the following:

(i) the highest numbered stereocenter on the open-chain defines whether the monosaccharide is in the D- or L-series.

(ii) the functional groups in the glucose

(iii) the equilibrium between the open-chain and the cyclic forms

(iv) conversion of the open chain to the cyclic form produces an additional stereocenter

(v) the position (i. e., the atom number) of the reacting groups in the open-chain form and the reactive group in the products

(vi) the orientation of the groups on C-1 in the cyclic form

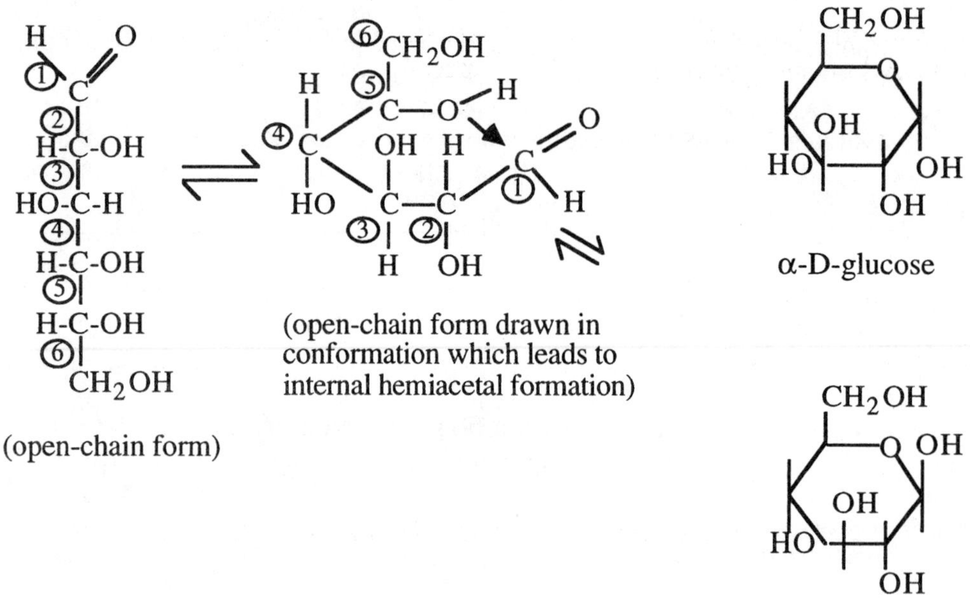

(open-chain form)

(open-chain form drawn in conformation which leads to internal hemiacetal formation)

α-D-glucose

β-D-glucose

Note that the anomeric carbon is at the C-1 position in the pyranose ring. As an exercise, draw the structures for D-fructose in an analogous reaction involving the formation of a five-membered ring containing a hemiacetal unit (formed from a ketone).

Two reactions to review are: (1) the oxidation of mono- or disaccharides that contain a hemiacetal unit and (ii) the hydrolysis reaction of the acetal linkages in polysaccharides that produces monosaccharides. The hydrolysis reaction is catalyzed by acids, or by enzymes in the living systems.

SELF-TEST QUESTIONS

<u>MULTIPLE CHOICE</u>. Select the correct answer from the choices listed. In some cases, more than one answer is correct.

1. Which of the following properties are associated with D-fructose?

   a. optically active
   b. undergoes mutarotation
   c. disaccharide
   d. is a ketose
   e. is or contains a reducing sugar
   f. contains an α(1--->4) linkage

2. Indicate which of the following properties is associated with maltose.
   a. rotates plane-polarized light
   b. contains a furanose ring
   c. contains a glycosidic bond
   d. is not a reducing sugar

3. Starch has which of the following characteristics?

a. β (1--->4) glycosidic linkages
b. made up of amylose and glycogen
c. is a monosaccharide
d. optically active

4. Which of the following molecules are disaccharides?

a. heparin
b. sucrose
c. maltose
d. fructose

5. Although the Haworth formula is a simple and useful representation of a cyclic sugar in many ways, it does not correctly represent the bond angles about the carbon atoms in the sugar molecule. These bond angles are:

a. 120°
b. 109.5^o
c. 150°
d. 90°

6. The oxidation of D-glucose by glucose oxidase to produce D-gluconic acid occurs at which carbon atom?

a. C-1
b. C-2
c. C-6
d. C-4

7. The IUPAC name for linear form of D-ribose , an aldopentose, is

a. 1, 2, 3, 4-Tetrahydroxypentanol
b. 2, 3, 4, 5- Tetrahydroxypentanal
c. 1, 2, 3, 4, 5-Pentahydroxypentanal
d. 1-Tetrahydroxypentanal

8. D-glucose can be converted to L-ascorbic acid by some plants and animals, but not humans. L-Ascorbic acid (vitamin C) contains a number of functional groups, but does not contain a (an)

a. lactone
b. chiral carbon
c. aldehyde
d. alcohol functional group

9. Indicate which groups an acidic polysaccharides contains.

a. phosphate
b. sulfate
c. acetate
d. carboxyl

10. Indicate the number of stereocenters that the 5-carbon sugar, D-ribose, contains. Refer to Table 19.1 in the text.

a. 2
b. 3
c. 4
d. 5

11. The carbon in the secondary alcohol group in D-glucose is carbon number:

a. 1
b. 3
c. 4
d. 6

12. The structure for D-gulose is shown below. Indicate which characteristics D-gulose exhibits.

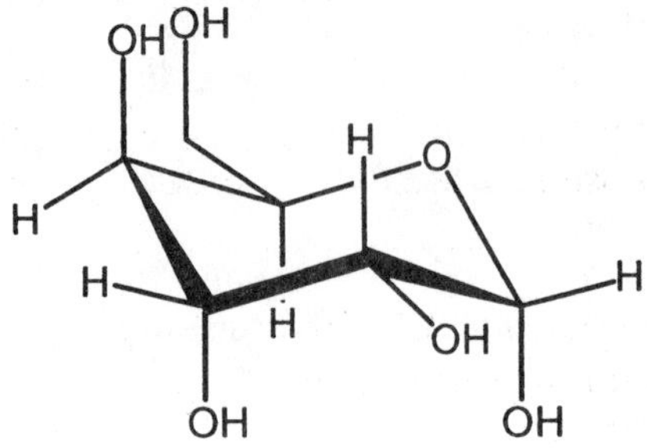

a. an aldopentose
b. a reducing sugar
c. shown in the α-configuration
d. is an acetal

<u>COMPLETION.</u> Complete the following statements with the appropriate word, phrase or number.

1. The process in which the α- and β-ring forms of a monosaccharide interconvert by way of the open chain structure is called ________________.

2. Fructose is a _______ sugar, with a ketone functional group at the 2-carbon. The oxidation of a fructose occurs because in basic solution, it is in equilibrium with the _________ form. An __________, which contains a carbon-carbon double bond, serves as the key intermediate for this equilibrium and the oxidation.

3. The hydrolysis of sucrose produces _______________ and ________________.

4. The polysaccharide which contains the β(1-->4) glycosidic linkage is ____________.

5. The Haworth projection of glucose does not indicate that the glucose is actually in a ___________ conformation, with the hydrogens or alcohol groups directed in either an __________ or an _____________ projection.

6. The ________ (type of reaction) of a monosaccharide yields a product generally called an _________. This reaction specifically converts glucose to _______ or _________ (two alternative names).

7. Amylopectin differs from amylose in that it contains ________________ that involve _______________ glycosidic linkages.

8. Blood type _______ is the universal donor because it has ____________________ of the cell-surface bound carbohydrate _______________ contained in A or B.

9. The acidic polysaccharide, _________, is naturally found in many human tissues. It is best known for its _________ activity in blood.

10. Lactose is an example of a ______________ disaccharide because it contains a _______________ unit. Therefore, it (<u>will or will not</u>) undergo mutarotation.

11. Two common features of heparin and hyaluronic acid are that they both contain __________ (rings) and _______ side groups.

12. Sugars are very soluble in ________ because the _____ groups on the sugar engage in ______ ______ interactions with water.

13. The number one carbon in glucose is the called the _________ carbon. It is this position that determines whether the D-glucose is in the _____- or the ____- form. These two forms of glucose are not enantiomers, but are _____________.

14. __________ is the name of the disaccharide that contains only _____ glucose residues, in which a carbon on each glucose is linked together by an _____________ (a functional group) linkage. This glycosidic bond is between carbon number ______ on one glucose and carbon number ______ on the other glucose. In the reaction of any two monosaccharides to yield a disaccharide, one molecule of ________ is also produced.

15. Humans cannot digest cellulose because they do not have the enzyme, ____________, which cleaves a ___________ linkage in cellulose.

ANSWERS TO SELF-TEST QUESTIONS

Multiple Choice

1. a, b, d, e
2. a, c
3. d
4 b, c
5. b
6. a
7. b
8. c
9. b, d
10. b
11. d
12. b, c

Completion

1. mutarotation
2. ketose, aldose, enediol
3. glucose, fructose
4. cellulose
5. chair, axial, equatorial
6. reduction, alditol, D-glucitol, D-sorbitol
7. chain branching, (1-6)
8. 0, neither, antigens
9. heparin, anticoagulant
10. reducing, hemiacetal, will
11. pyranose, acidic
12. water, -OH, hydrogen bonding
13. anomeric, α–, β–, diastereomer
14. Maltose, 2, ether, 1, 4, water
15. β-glucosidases, β(1-4)

It takes a membrane to make sense
out of disorder in biology.
- Lewis Thomas
The Lives of a Cell

20

Lipids

CHAPTER OBJECTIVES

Lipids are defined **operationally** - that is, defined by their properties and not their composition and structure. Molecules in a cell are collectively considered as lipids if they can be extracted from the cell extract with non-polar solvents. This classification procedure is not very rigorous and as a result a rather disparate melange of molecules, with very different compositions, structures and functions are enveloped within the definition. The focus of this chapter is therefore to define the (1) **COMPOSITION**, (2) **STRUCTURE**, (3) **PROPERTIES** and (4) **FUNCTION** of these cellular molecules. Keep this in mind as you review the material. As you will see, the functions range from providing the basic structure of all membranes, to highly specific hormones, some of which regulate metabolism, while others are very specific and define our sex.

After you have studied the chapter and worked the assigned exercises in the text and the study guide, you should be able to do the following.

1. Name the three major roles of lipids in human biochemistry.

2. List the four main subclasses of lipids and give a few examples of each.

3. Draw the general structure of a mono-, di- and triglyceride (all fats) which includes saturated and/or unsaturated fatty acids.

4. Explain how the unsaturation in fatty acids influences their melting point.

5. Write out the general equations for the (1) hydrogenation of unsaturated fats and oils and for the (2) saponification of all fats or oils.

6. Define the lipid bilayer and describe the essential features of the fluid mosaic model of a membrane.

7. Distinguish between transmembrane transport processes that take place by passive, facilitated and active transport mechanisms.

8. Draw a phosphoglyceride and a sphingolipid and indicate their main function.

9. Draw the structure for cholesterol and outline the relationship between cholesterol, progesterone, the adrenocorticoid hormones, and the sex hormones.

10. Characterize the four classes of lipoproteins.

11. Outline the interrelationship between the HDLs, LDLs, cholesterol and its metabolism, and the important roles that each plays in the development of arteriosclerosis.

12. Outline the details of how bile salts and lipases work in concert to break down lipids.

13. Draw the structure of arachidonic acid and point out some common structural changes that occur on production of prostaglandins, leukotrienes and thromboxanes.

14. Discuss the function of cyclooxygenases, COX-1 and COX-2, in human physiology and the influence of nonsteroidal anti-inflammatory drugs (NSAIDs) on these critical enzymes.

15. Define the important terms and comparisons in this chapter and give specific examples where appropriate.

IMPORTANT TERMS AND COMPARISONS

Lipids
Saturated and Unsaturated Fatty Acids
Essential Fatty Acids
Hydrogenation
Soaps, Detergents, and Micelles
Phospholipids and Glycolipids
Membranes and Lipid Bilayer
Passive, Facilitated and Active Transport
Cephalin and Lecithin
Glycerol
Multiple Sclerosis and the Myelin Sheath
Lipid Storage Disease and Tay-Sachs
Lipoproteins
LDL-Receptors
COX-1 and COX-2 Enzymes
Anabolic Steroids
Mineralcorticoids and Glucocorticoids
Sex Hormones - Testosterone and Estradiol
Progesterone and RU486
COX-1 and COX-2 Enzymes
Lipase

Mono-, Di- and Triglycerides
Fats, Oils and Waxes
Polyunsaturated Oils
Saponification
Hydrophobic and Hydrophilic
Glycerophospholipids and Sphingolipids
Phosphatidylcholine or Lecithin
Transmembrane Channels
Phosphatidylinositol 4, 5-Bisphosphate (PIP2)
Sphingolipids - Sphingosine and Sphingomyelin
Glycolipids - Cerebrosides and Gangliosides
Steroid and Cholesterol
HDL, LDL, VLDL and Chylomicrons
HMG-CoA Reductase and Statin Drugs
Progesterone
Sex Hormones and Adrenocorticoid Hormones
Aldosterone and Cortisol
Oral Contraceptives
Bile Salts
Prostaglandins, Leukotrienes and Thromboxanes
Steroid and **N**on**s**teroidal **A**nti-**I**nflammatory **D**rugs (NSAIDS)

SELF-TEST QUESTIONS

MULTIPLE CHOICE. In the following exercises, select the correct answer from the choices listed. In some cases, two or more choices will be correct.

1. Fats differ from waxes in that fats have:

   a. more unsaturation
   b. a glycerol backbone
   c. higher melting points
   d. longer fatty acids

2. Oleic acid is an unsaturated fatty acid containing:

   a. 12 carbons
   b. 14 carbons
   c. 16 carbons
   d. 18 carbons

3. Both stearic acid and linoleic acid contain 18 carbons. However, linoleic acid is unsaturated, while stearic acid is saturated. The melting point of stearic acid:

   a. is higher
   b. is lower
   c. is the same as linoleic acid
   d. relative to that for linoleic acid cannot be predicted

4. Liquid soaps are generally salts of fatty acids in which the cation is:

   a. K^+
   b. Na^+
   c. Ca^{+2}
   d. Mg^{+2}

5. Indicate which of the following structures represents cholesterol?

a.

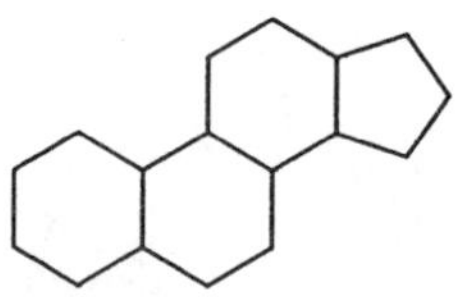

b.

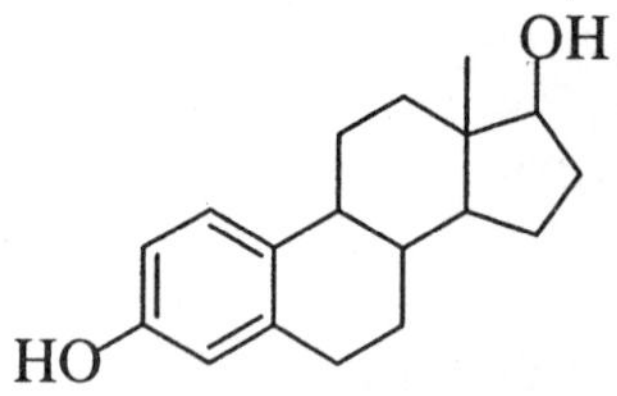

c.

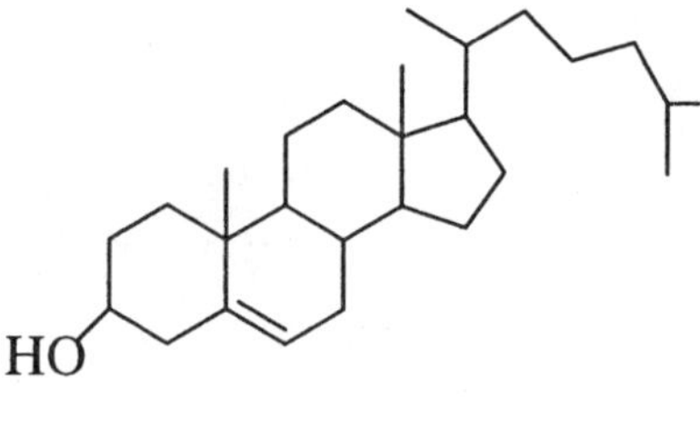

d.

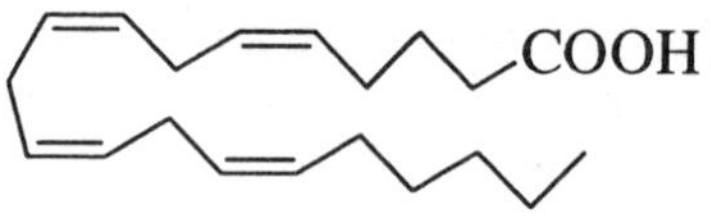

6. Prostaglandin group, PGE_2, possesses:

   a. ring closure at C-8 and C-12
   b. 20 carbon atoms
   c. a carbonyl group at C-9
   d. two olefinic (carbon-carbon) double bonds

7. Multiple sclerosis is thought to be associated with:

   a. degradation of the myelin sheath
   b. inhibited prostaglandin synthesis
   c. lipid storage
   d. a deficiency of enzyme COX-1

8. Lipid bilayers are composed mainly of:

   a. complex lipids
   b. steroids
   c. prostaglandins
   d. fatty acids

9. Basic hydrolysis or saponification of triglycerides will produce:

   a. glycerol
   b. fatty acids
   c. salts of fatty acids
   d. bile salts

10. The 18-carbon backbone of sphingosine is found in:

    a. sphingomyelin
    b. myelin sheath
    c. phosphatidylcholine
    d. testosterone

11. Oils are:

    a. liquid fats
    b. completely saturated
    c. high molecular weight fatty acids
    d. colored

12. Derivatives of what molecules are effective oral contraceptives?

    a. progesterone
    b. estradiol
    c. PGF_{2a}
    d. cortisone

13. Low density lipoprotein (LDL) has which of the following characteristics:

    a. Contains more cholesterol than high density lipoprotein (HDL).
    b. Soluble in the blood plasma
    c. Carries cholesterol to the cells
    d. Binds to receptor molecules on the cell surface within the coated pits.

14. Anabolic steroids are

    a. powerful female sex hormones
    b. very effective detergents that aid digestion
    c. effective in the treatment of head aches
    d. used by athletes to increase muscle development

15. The fluid mosaic model of the membrane is characterized by

    a. a lipid bilayer made up of complex lipids
    b. proteins embedded in the bilayer
    c. cholesterol molecules inserted in the bilayer
    d. all of these characteristics

16. Derivatives of this group of membrane lipids function as signaling molecules in chemical communication.

    a. cephalins
    b. ceramides
    c. phosphatidyl inositol
    d. glucocerebrosides

17. The sex hormone that contains an aromatic ring is:

    a. progesterone
    b. testosterone
    c. estrogen
    d. aldosterone

18. Indicate which of the following characteristics bile salts exhibit.

    a. contain an aromatic ring
    b. derived from the oxidation of cholesterol
    c. provides a mechanism to eliminate excess cholesterol and its breakdown products
    d. contain either a glycine or taurine group

<u>COMPLETION</u>. Write the word, phrase or number in the blank space in answering the question.

1. Energy is stored in our bodies in the form of ______________ because burning it produces twice as much energy as the burning of ______________.
2. Aspirin has been found to inhibit ______________ synthesis.
3. Both the physiologically potent prostaglandins and leukotrienes are derived from the 20-carbon, unsaturated fatty acid called __________ ________.
4. The enzymes involved in lipid metabolism are called ______________.
5. A widely known disease associated with lipid storage is ______________.
6. ______________ are negatively charged salts, which are powerful detergents used in the digestion of lipids. They are derived from the oxidation of ______________.
7. The hydrophobic segment of a detergent is associated with the ______________.
8. All unsaturated fatty acids exist as ______________ at room temperature.
9. Rancidity is caused in part by the ______________ of the fat or oil to aldehydes or by the production of ______________ by the hydrolysis of triglycerides.
10. The fluid mosaic model of membranes proposes that ______________ molecules are floating on and within the ______________.
11. Fats are chemically the equivalent of an ___________ (a functional group), composed of ___________ and _____ _____ with an even number of carbons. Waxes are similar, but are virtually always ________ (liquid or solid) because the alcohol component is a _______ _______ alcohol.
12. The common name for phosphatidylcholine is ______________.
13. Leukotrienes are ______________ of hormonal responses. Although similar in structure

to prostaglandins, there is no _________ _________ in the structure of leukotrienes.

14. A faulty cholesterol transport system is often characterized by high levels of ________ and a relatively low level of ___________. In the disease familial hypercholesterolemia., this results from not enough functional. ______ _______ (two words).

15. Aldosterone is a ________________, which regulates the concentrations of _____ and is synthesized in the ________________ gland.

16. The target molecule for most statin drugs, which are used to inhibit cholesterol synthesis in the liver, is the enzyme, _______________.

17. I was once told that the way to remember the steroid skeleton is to compare it to a house that contains "three rooms and a bath". In this analogy, the three rooms refer to the three _________ rings and the bath to the __________ ring.

ANSWERS TO SELF-TEST QUESTIONS

Multiple Choice

1. b
2. d
3. a
4. a
5. c
6. a, b, c, d
7. a
8. a
9. a, c
10. a
11. a
12. a
13. a, b, c, d
14. d
15. d
16. c
17. c
18. b, c, d

Completion

1. fat, carbohydrates
2. prostaglandin
3. arachidonic acid
4. lipases
5. any of the 5 listed in Box 20.F in the text, such as Gaucher's disease
6. Bile salts, cholesterol
7. non-polar hydrocarbon chain
8. liquids
9. oxidation, fatty acids
10. protein, lipid bilayer
11. esters, glycerol, fatty acids, solid, long chain
12. lecithin
13. mediators, ring closure
14. LDL, HDLs, LDL receptors
15. mineralcorticoid, ions, adrenal
16. HMG-CoA reductase
17. 6-membered, 5-membered

"Isn't it astonishing that all these secrets
have been preserved for so many years
just so that we could discover them."
- Orville Wright
Miracle at Kitty Hawk

21

Proteins

CHAPTER OBJECTIVES

This chapter, similar to the previous chapter on lipids, concentrates on the (1) **COMPOSITION**, (2) **SEQUENCE**, (3) the **levels of STRUCTURE** and the (4) diverse **FUNCTIONS** of proteins. Proteins and/or enzymes are associated with, in some way, virtually all reactions in the cell and therefore will be of interest in all aspects of cellular regulation and metabolic processes.

After you have studied the chapter and worked the assigned exercises in the text and the study guide, you should be able to do the following.

1. List the eight major functions of proteins and give an example of a real protein that is involved in each biological function.

2. Write out the structures for at least two amino acids which are representative of an amino acid with a (1) nonpolar, (2) polar, but neutral, (3) acidic and (4) basic R group. Also indicate the unique character of the amino acids glycine, proline and cysteine.

3. Demonstrate the amphiprotic character of amino acids by writing the reactions for the zwitterion with an (1) acid or (2) base.

4. Write out the reversible reaction for the oxidation of cysteine and indicate how the production of cystine in a **protein** might stabilize the tertiary structure of proteins.

5. Draw the structure for a tripeptide, pointing out (1) the peptide linkages, (2) any stereocenters in the backbone, (3) the N- and C-terminal amino acid residues and (4) the R groups.

6. Name the four levels of structure in proteins and briefly describe what each is associated with.

7. Indicate the four forces that are involved in determining the final protein structure and draw a representation of each.

8. Characterize the secondary structures in α-helix and β-pleated sheet in terms of structure and the types of forces involved in each.

9. Indicate what changes occur in proteins by the addition of specific denaturation agents and list some practical applications of denaturation.

10. Define the important terms and comparisons in this chapter and give specific examples where appropriate.

IMPORTANT TERMS AND COMPARISONS

Protein and Peptide
Enzyme
Hydrophobic and Hydrophilic Side Chains
Isoelectric Point, pI
Amino Acid Names, 3-Letter- and 1-Letter Abbrev.
Cysteine and Cystine
Hypoglycemia
Fibrous and Globular Proteins
Di-, Tri-, Polypeptide and Proteins
Glycation, AGE, Diabetes and Aging
Primary, Secondary and Tertiary Structure
Salt Bridges and Hydrophobic Interactions
Intra- and Intermolecular Hydrogen Bonding
Prosthetic Group
Hemoglobin
Homo- and Heterozygote
Glycoproteins
Denaturation
Proteomics

Catalyst
Alpha Amino Acid and Amino Acid Residue
Zwitterion
Nonpolar, Polar, Acidic & Basic Side Chains
Amphiprotic Molecules
Thiol or Sulfhydryl (-S-H) and a Disulfide Bond (-S-S-)
N-Terminal and C-Terminal Amino Acids
Peptide Backbone
Prion and Pathological Conditions
Quaternary Structure
α-Helix, β-Pleated Sheet and Random Coil
Triple Helix Structure of Collagen
Tropocollagen and Collagen
Sickle Cell Anemia
Alpha Chains in Globular Hemoglobin
O-Linked and N-Linked Saccharides
Integral Membrane Proteins
Chaperone Proteins

SELF-TEST QUESTIONS

MULTIPLE CHOICE. In the following exercises, select the correct answer from the choices listed. In some cases, two or more choices will be correct.

1. The number of tetrapeptides that can be made from any of the 20 amino acids is:

   a. 1.6×10^5
   b. 20
   c. 4
   d. 80

2. Glycine is an unique amino acid because it:

   a. has no enantiomer
   b. has a sulfur containing R group
   c. cannot form a peptide bond
   d. is an essential amino acid

3. The α-helix is a common form of:

a. primary structure
b. secondary structure
c. tertiary structure
d. a denatured region

4. Which of the following is a transport protein?

a. hemoglobin
b. pepsin
c. collagen
d. oxytocin

5. A zwitterion has which of the following characteristics?

a. a high melting point
b. no net charge
c. soluble in water
d. all of above

6. Arginine is a basic amino acid while alanine is nonpolar. The isoelectric point for arginine is:

a. higher than for alanine
b. lower than for alanine
c. about the same as alanine
d. cannot predict

7. Hemoglobin is an oxygen carrying protein made up of two alpha chains and two beta chains. Hemoglobin has:

a. 2 subunits
b. 1 subunit
c. no subunits
d. 4 subunits

8. Which of the following R groups associated with an amino acid can take part in hydrogen bonding?

a. $-CH_2SCH_3$ (methionine)
b. $-CH_2OH$ (serine)
c. $-CH_2CO_2H$ (aspartic acid)
d. $-CH_2CH_2CH_2CH_2NH_3^+$ (lysine)

9. Hydrophobic interactions occur between the R groups with:

a. acidic character
b. nonpolar character
c. basic character
d. all of above

10. The number of amino acid residues in the cyclic nonapeptide, vasopressin, is:

a. 7
b. 9
c. 19
d. 90

11. An example of a globular protein that is a hormone is

a. hemoglobin
b. keratin
c. collagen
d. insulin

12. The secondary structure in an alpha helix is stabilized (held together) by

a. intramolecular hydrogen bonds
b. hydrophobic interactions
c. electrostatic bonds
d. intermolecular hydrogen bonds

13. An example of a prosthetic group is

a. methionine
b. the heme group
c. an amide group
d. any functional group

14. Muscle proteins that are involved in movement are:

a actin and myosin
b. heme and hemoglobin
c. collagen and protocollagen
d. fibrinogen and platelets

15. Two amino acid residues **in a protein** that can take part in electrostatic interactions are:

a. alanine and isoleucine
b. histidine and lysine
c. methionine and cysteine
d. glutamic acid and phenylalanine

16. The charge on the tripeptide, leu-arg-lys, at pH 7, is:

a. +1
b. +2
c. +3
d. zero

17. The ingestion of heavy metals, such as lead (Pb^{+2}) or mercury (Hg^{+2}) can be poisonous because they denature proteins by binding to:

a. methionine
b. cysteine
c. arginine
d. tyrosine

COMPLETION. Write the word, phrase or number in the blank space in answering the question.

1. In the tripeptide drawn below, circle and clearly label the following units: peptide linkages; C- and N-terminal end; R groups.

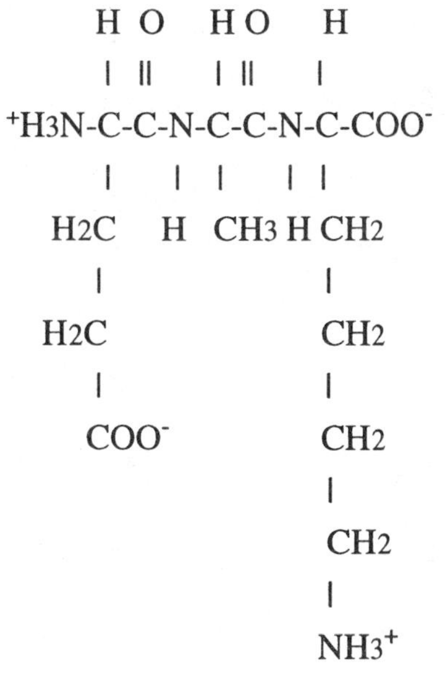

2. Tropocollagen is the _____________ form of collagen and is made up of ___________ _______________ units. Insoluble collagen results from covalent _______________ of two _______________ residues on adjacent chains in the helix.

3. A covalent bond in the __________ ___________ (protein backbone), is the force which is associated with the _______________ structure of a protein.

4. Agents that can reversibly denature a single subunit protein do this by breaking forces involved in the _______________ and _______________ (levels of) structure.

5. Sickle cell trait occurs in people who have _______________ gene(s) programmed to produce the protein in sickle cell hemoglobin.

6. The α-helix is stabilized by intramolecular hydrogen bonding between _______ amino acid residues along the peptide chain, while the hydrogen bonding in forming a β-pleated sheet structure is stabilized by intramolecular hydrogen bonding between ________ amino acid residues.The interaction,in both cases, is between the _______________ and the _______________ groups in the peptide __________.

7. N-Linked saccharides have the saccharide linked directly to ___________, while O-linked saccahrides may be linked to either __________, __________ or _________.

8. Complete the following reactions:

a.
```
      H
   +  |
   H3N-C-CO2-  +  H+  ----->      ____________________
      |
      R
```

b.
```
        H           [O]
     +  |          ----->
2  H3N-C-CO2-      <-----         ____________________
        |           [H]
        CH2
        |
        SH
```

c.
```
                        heat
tetrapeptide  +  HCl   ------->   ____________________
```

d.

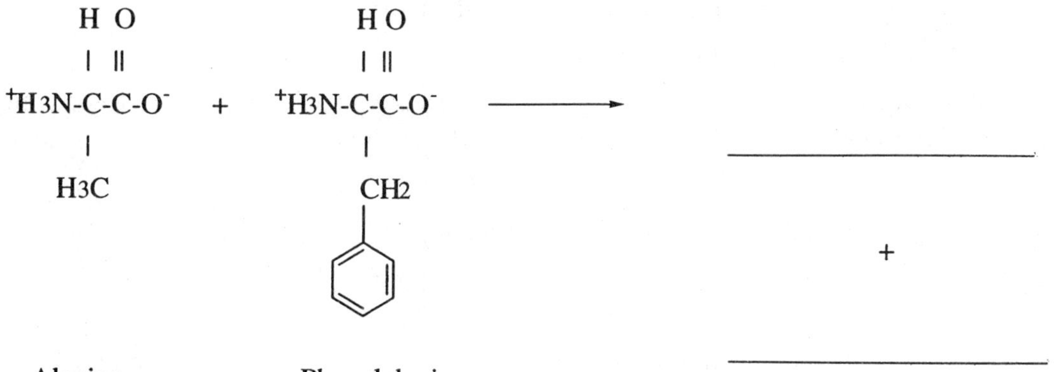

Alanine Phenylalanine

9. _______________ structure occurs only in proteins that have _______________ subunits, such as in hemoglobin.

10. The charge on hydrophobic amino acid, alanine, at very low pH is ________, while at high pH it is __________.

ANSWERS TO SELF-TEST QUESTIONS

Multiple Choice

1. a
2. a
3. b
4. a
5. d
6. a
7. d
8. b, c, d
9. b
10. b
11. d
12. a
13. b
14. a
15. b
16. b
17. b

Completion

1.

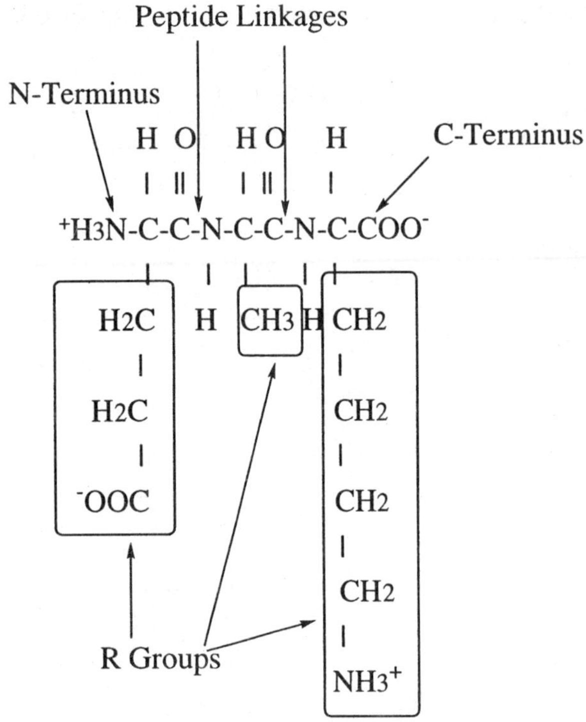

2. soluble; triple helical; crosslinking; lysine
3. peptide linkage; primary
4. secondary; tertiary
5. one
6. adjacent, non-adjacent, C=O, H-N, backbone
7. asparagine, serine, threonine, hydroxylysine
8.

a. 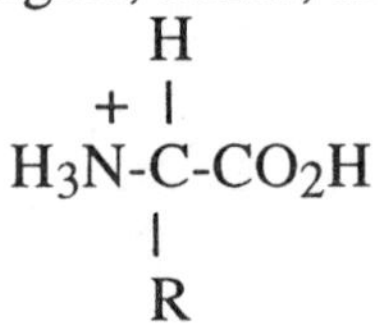

b. 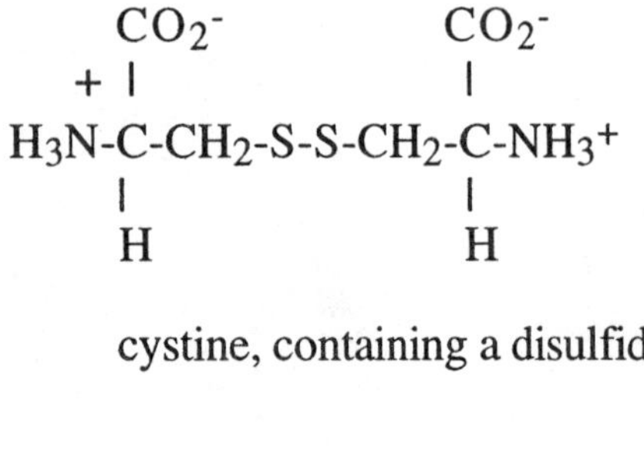

cystine, containing a disulfide bond

c. an acid hydrolysis reaction to yield 4 amino acids

d. Two different dipeptides are produced with the same composition, but with different sequence. In theory, the "homodipeptides", ala-ala and phe-phe, may also be formed.

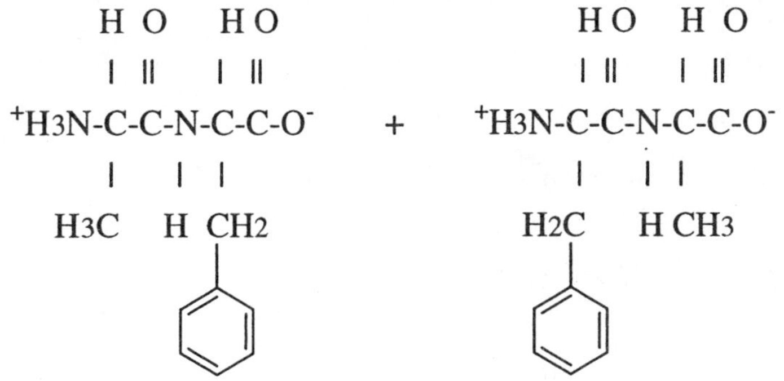

9. Quaternary, multiple (2 or more)
10. +1, -1

Were it not equipped with catalysts, every
living unit would be a static system
- F.G. Hopkins
Science, 78, 219 (1933)

22

Enzymes

CHAPTER OBJECTIVES

After you have studied the chapter and worked the assigned exercises in the text and the study guide, you should be able to do the following.

1. Explain the statement that enzymes increase the rate of the reaction, but do not change the position of the equilibrium.

2. List the six major categories of enzymes and indicate the general type of reaction that the enzyme catalyzes.

3. Distinguish between the terms, enzyme, cofactor, coenzyme, apoenzyme, proenzyme, isoenzyme and active site and regulatory site.

4. Describe, in both words and in figures, the effect the following have on the rate of a catalyzed reaction:
   - enzyme concentration
   - substrate concentration
   - temperature
   - pH
   - enzyme inhibitors

5. Outline and compare the proposals in the (1) lock and key model and the (2) induced fit model to explain enzymatic action and specificity.

6. Describe how competitive and a non-competitive reversible inhibitors produce their effect.

7. Define active site and list factors associated with it which are responsible for its specificity toward molecular substrates.

8. Explain how the activity of an allosteric enzyme can be modulated in a positive or negative way by regulator interactions at the regulatory site. Relate this effect to the feedback control mechanisms found in multiple-step reaction schemes.

9. Describe how sulfa drugs can be used as competitive inhibitors to kill harmful bacteria in our bodies, but have no effect on our metabolism.

10. Define the important terms and comparisons in this chapter and give specific examples where appropriate.

IMPORTANT TERMS AND COMPARISONS

Enzyme and Catalyst
Substrate and Competitive Inhibitor
-ase
Substrate and Active Site
Proenzyme and Zymogens
Reversible and Irreversible Inhibition
Oxidoreductase
Transferase
Hydrolase
Lyase
Isomerase
Ligase
Active Site and Regulator Site
Regulatory or Allosteric Enzymes
Regulator and Regulator Site
Feedback Control of Enzyme Activity
Enzyme Modification
Ribozymes
Cofactor and Coenzyme
Apoenzyme and Holoenzyme
Enzyme-Substrate Complex
Activation and Inhibition
Competitive and Non-Competitive Inhibitors
pH Profile for Enzymatic Activity
Lock-in Key and Induced-Fit Models
Irreversible Inhibitor
Sulfa Drugs and Para-Aminobenzoic Acid
Le Chatelier's Principle
Allosterism and Allosteric Enzymes
Positive & Negative Modulation of Regulatory Enzyme Activity
Isozymes or Isoenzymes
Angiotensin-Converting Enzyme (ACE)
Kinases

SELF-TEST QUESTIONS

MULTIPLE CHOICE. In the following exercises, select the correct answer from the choices listed. In some cases, two or more choices will be correct.

1. A molecule that is structurally similar to the substrate for an enzyme will probably be a:

   a. competitive inhibitor
   b. cofactor
   c. regulator
   d. noncompetitive inhibitor

2. The site on an allosteric enzyme that is directly involved in modulation is called

   a. the prosthetic group
   b. regulatory site
   c. active site
   d. target site

3. The multistep sequence of reactions shown below uses a regulatory mechanism to control the amount of D produced. In these general schemes, the last product, D, in the reaction sequence is involved in a feedback mechanism of inhibition. Molecule D in the sequence usually

   a. inhibits E_1
   b. inhibits E_2
   c. inhibits E_3
   d. inhibits all enzymes

$$A \xrightarrow{E_1} B \xrightarrow{E_2} C \xrightarrow{E_3} D$$

4. Competitive inhibition can be overcome by:

   a. increasing substrate concentration
   b. increasing pH
   c. decreasing temperature
   d. all of above

5. The pH influences most enzymatic reactions because pH changes can:

   a. completely hydrolyze the protein
   b. produce protonation or deprotonation of essential amino acid residues in the active site
   c. change its primary structure
   d. effect the optical activity of the protein

6. A significant increase in the temperature (as when a patient has a very high fever) of an enzymatic reaction will reduce the catalytic rate because the:

   a. enzyme acts on itself as an inhibitor
   b. protein is partially or completely denatured
   c. protein undergoes hydrolysis
   d. heat cats as a competitive inhibitor

7. Consider the possible shapes that a molecule might have to act as a competitive inhibitor to the reaction shown below:

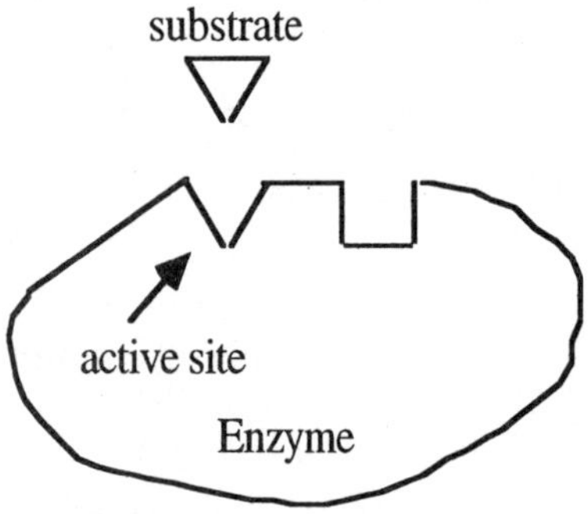

Indicate which of the following molecules (molecular shapes) could potentially act as a competitive inhibitor.

a.

c.

b.

d.

8. The apoprotein corresponds to which part of an enzyme?

a. coenzyme
b. protein portion
c. cofactor
d. part cleaved off in a proenzyme

9. The site on the enzyme at which the substrate interacts is called the

a. regulatory site
b. modulator site
c. active site
d. allosteric site

10. A kinase is an enzyme that acts on another enzyme, A, to change its activity. A kinase modifies enzyme A by

a. cleaving enzyme A to produce a smaller active enzyme
b. adding a missing coenzyme in the reaction
c. modifies enzyme A by adding a phosphate group on the enzyme
d. modifies enzyme A by adding a glycosyl group on the enzyme

11. Indicate which of the following characteristics an allosteric enzyme exhibits.

a. A regulator molecule can bind strongly to the active site.
b. The binding of a regulatory molecule to the regulatory site always inhibits enzymatic activity.
c. The binding of the regulatory molecule to the regulatory site can either increase or decrease the enzymatic activity.
d. The active site and the regulatory sites are always in different locations and cannot be identical.
e. In most cases, the allosteric enzyme is a multisubunit protein.

12. Characteristics of a proenzyme are

a. they are made in the cell in an inactive form
b. they do not contain all the amino acid residues for a complete and active enzyme
c. they are usually cleaved (cut) to produce a smaller protein that becomes the active enzyme
d. they do not have the correct prosthetic group for activity

13. The protein modification that is most common in the activation of enzymes is

a. methylation
b. phosphorylation
c. addition of an aromatic group
d. addition of an ester group

14. The enzyme, prostaglandin enderoperoxide synthase (PGHS), (Box 22G) is <u>unusual</u> in that it

a. has two active sites and catalyzes two different reactions
b. has two regulatory sites and no active site
c. can be inhibited by common over-the-counter-drugs
d. is primarily composed of a polysaccharide

15. The small molecule, urea, is used to denature proteins. Its structure is $H_2N\text{-}\underset{\underset{O}{\|}}{C}\text{-}NH_2$
Its ability to denature proteins is primarily because

a. it cleaves the peptide backbone
b. it breaks up essential hydrogen bonding in the secondary and tertiary structure
c. it breaks up hydrophobic interactions in the secondary and tertiary structure
d. it reacts with the hydroxyl group on glycine residues

16. Isoenzymes are enzymes that

a. occur in different forms in different tissues and catalyze the same reaction
b. are always composed of only one subunit
c. must be modified by a kinase to become active
d. are unaffected by extreme temperatures

<u>Completion</u>. Write the word, phrase or number in the blank space in answering the question.

1. Classify each of the following enzymes into one of the six major classes.

<u>Enzyme</u>	**<u>Class</u>**
a. Acid Phosphatase	
b. Alcohol Dehydrogenase	
c. DNA (Topo)isomerase	
d. Pepsin	
e. Tyrosine-tRNA Synthetase	
f. Pyruvate Kinase	
g. Acetylcholinesterase	
h. Aspartate Amino Transferase	

2. Lipases are enzymes that are (<u>more or less</u>) specific than ACE.

3. Of the two main models used to explain enzymatic action, the ________________ model is regarded as more flexible than the ________________ model.

4. A number of enzymes require a non-protein component called a ________________ or ________________ to exhibit enzymatic activity.

5. Trypsinogen is an example of a ________________ that can only be ________________ by removal of a portion of the protein.

6. If the enzyme concentration is tripled, the reaction rate will ____________.
(Refer to Figure 22.3 in the text).

7. Heavy metals, such as Hg^{+2} or Pb^{+2}, act as ________________ inhibitors to many enzymes.

8. Label the following plot to show a pH profile for an enzyme in which the greatest activity occurs at pH 4.5 and decreases to nearly zero at pH 7.0 and 2.0. Make the plot with enzyme rate on the vertical axis and pH on the horizonal axis.

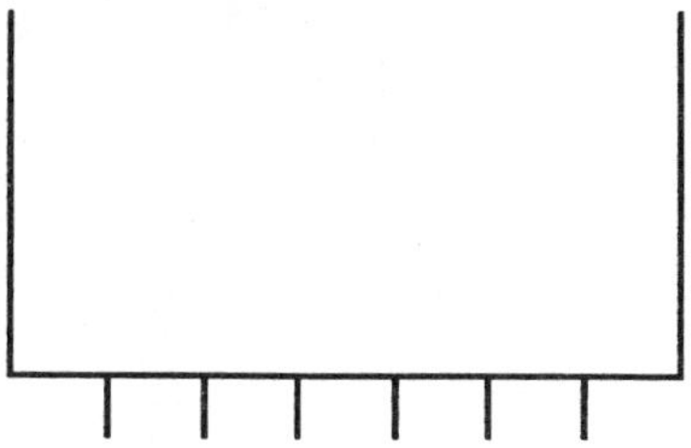

9. Draw the substrate, enzyme and enzyme-substrate (E-S) complex as they may be viewed in the induced fit model.

__________		__________	--------> <--------	____________________
Substrate	+	Enzyme		(E-S) complex

10. Although cofactors can be either _________ or ___________, coenzymes are always _________ and are synthesized from a vitamin.

ANSWERS TO SELF-TEST QUESTIONS

Multiple Choice

1. a
2. b
3. a
4. a
5. b
6. b
7. a
8. b
9. c
10. c
11. c, d, e
12. a, c
13. b
14. a
15. b
16. a

Completion

1. a. Hydrolase
   b. Oxidoreductase
   c. Isomerase
   d. Hydrolase
   e. Ligase
   f. Transferase (specific for phosphates)
   g. Hydrolase
   h. Transferase
2. less
3. induced fit; lock and key
4. cofactor, coenzyme
5. proenzyme, activated
6. triple
7. irreversible

8.

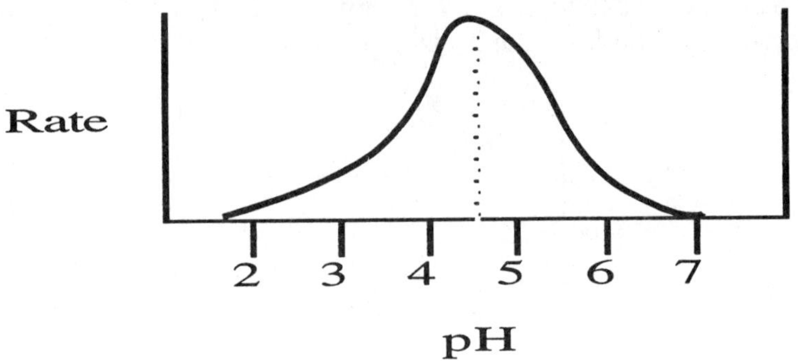

9. The (free) enzyme structure (without substrate bound to it) is different than when the substrate is bound to the active site.

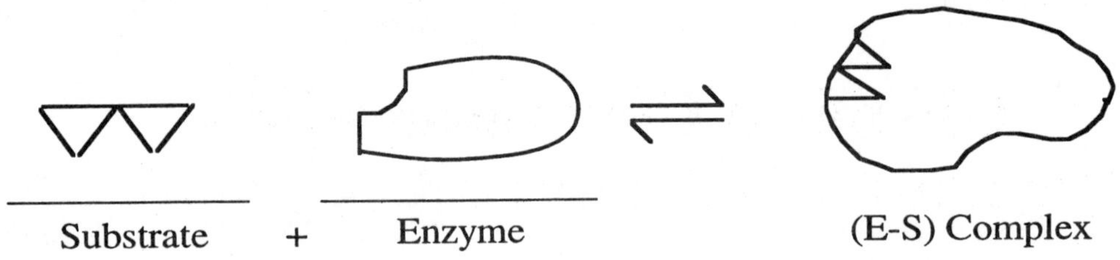

10. inorganic ions (such as Mg^{+2}, Zn^{+2}, etc). organic compounds, organic compounds

One, if by land, or two, if by sea.
- Paul Revere's Ride
Henry Wadsworth Longfellow

23

Chemical Communications: Neurotransmitters and Hormones

CHAPTER OBJECTIVES

After you have studied the chapter and worked the assigned exercises in the text and the study guide, you should be able to do the following.

1. Tabulate the two general types of chemicals involved in intercellular communication, indicating (1) their general make up, (2) what each acts on, (3) the time frame of their action, (4) the cells or organs which are involved in the communication and (5) whether secondary messengers are involved in the process.

2. Explain the role of the axons, dendrites, the synapse, vesicles and receptor sites in the neurotransmission by acetylcholine.

3. Describe how acetylcholine bound to postsynaptic receptor sites can be removed normally in the cell and how muscle relaxants or nerve gases can prevent this.

4. Discuss the action of drugs that function as agonists and antagonists on cellular receptors.

5. Discuss the role and mechanism of action of ligand-gated ion channels.

6. Outline the interrelationship between chemical messenger, membrane receptor, G-proteins and the enzyme, adenylate cyclase and the ultimate target.

7. Distinguish between the H_1 and H_2 histamine receptors in terms of their location, function and targeted treatment in disease.

8. Outline the mechanism of action and the final inactivation of adrenergic neurotransmitters by monoamine oxidases (MAO).

9. List the three types of hormones, their general make up and the types of action that each stimulates.

10. List a number of hormones, the gland that secretes the hormone and the specific action or function of the hormone (consult Table 23.2 in text).

11. Define the important terms and comparisons in this chapter and give specific examples where appropriate.

IMPORTANT TERMS AND COMPARISONS

Receptor, Chemical and Secondary Messengers
Agonists and Antagonists
Endocrine Gland and Hormones
Adenylate Cyclase and cAMP
Cholinergic, Amino Acid, Adrenergic And Peptidergic Neurotransmitters
Acetylcholine and Acetylcholinesterase
Primary and Secondary Messengers
Phosphodiesterase
Calmodulin
L-Dopa and Dopamine
Alzheimer's Disease, β-amyloid Plaques & Tau
Histamine and Histamine Receptors, $\mathbf{H_1}$ and $\mathbf{H_2}$
Transporters and Reuptake Process
Enkephalins
Histidine and Histamine
Diabetes Mellitus and Non-Insulin-Dependent Diabetes
Steroid Hormones
Nerve Gases and Sarin (Agent GB, Tabun)
Neurons, Axons, Dendrites
Neurotransmitters and Hormones
Synapse, Presynapse, Postsynapse and Vesicles
Adrenergic, Peptidergic and Steroid Hormones
Transmembrane Proteins
Ligand-Gated Ion Channel Receptors
G-Protein-cAMP Cascade and Signal Transduction
Parkinson's Disease and L-Dopa
Reversible and Irreversible Inhibitors of Acetylcholinesterase
Monoamine Oxidase (MAO)
Morphine
NO and NO Synthase
Helicobacter pylori Bacteria and Ulcers
Inhibitory Neurotransmitters
Tamoxifen and Estrogen Receptors
Five Classes of Chemical Messengers

SELF-TEST QUESTIONS

MULTIPLE CHOICE. In the following exercises, select the correct answer from the choices listed.

1. Decamethionium, which has a structure similar to that for acetylcholine, inhibits acetylcholinesterase. It is a

   a. non-competitive inhibitor
   b. chemical mediator
   c. competitive inhibitor
   d. toxic poison

2. Of the natural communication chemicals presented in the chapter, indicate the one which acts over the longest distance.

   a. neurotransmitters
   b. steroid hormones
   c. c-AMP
   d. NO

3. Acetylcholinesterase hydrolyzes which linkage in acetylcholine, $H_3C\overset{\overset{\displaystyle O}{||}}{C}OCH_2CH_2\overset{+}{N}(CH_3)_3$?

a. $H_3C \overset{\downarrow}{----} \underset{\underset{\displaystyle O}{||}}{C}$

b. $H_3C\underset{\underset{\displaystyle O}{||}}{C} \overset{\downarrow}{----} O$

c. $H_3C\underset{\underset{\displaystyle O}{||}}{C}O \overset{\downarrow}{----} C$

d. $H_3\underset{|}{C} \overset{\downarrow}{----} N(CH_3)_3$

4. $\mathbf{H_1}$ receptors found in the respiratory tract are blocked by the interaction of

a. insulin
b. L-DOPA
b. enkephalins
d. antihistamines

5. The second messenger in the action of monoamine neurotransmitters is in most cases

a. d-AMP
b. c-AMP
c. AMP
d. ATP

6. The administration of L-Dopa can reverse the symptoms of Parkinson's disease. Dopamine is produced from L-Dopa by a

a. hydrolysis reaction
b. oxidative decarboxylation reaction
c. decarboxylation reaction
d. oxidative reaction

7. Protein kinases function as

a. phosphate transfer enzymes
b. structural membrane proteins
c. secondary messengers
d. ion translocating proteins

8. Neurotransmitters are stored in vesicles in the

a. axon
b. dendrites
c. presynaptic site
d. postsynaptic site

9. Glucagon is an example of a

a. peptide hormone
b. steroid hormone
c. monoamine neurotransmitter
d. receptor

10. An agonist is a molecule that

a. stimulates antibody production
b. competitively inhibits acetylcholinesterase
c. activates receptor sites
d. reduces pain

11. Steroid hormones produce their effect in the cell by

a. binding to the membrane
b. activating a protein kinase
c. stimulating c-AMP production
d. activating one or more genes

12. The hormone classes include all the molecules listed below, with the exception of

a. peptides
b. nucleotides
c. steroids
d. small molecules, such as amino acids

13. Indicate which of the following are general characteristics found in ligand-gated ion channels (receptors).

a. They are multiple subunit proteins
b. They are transmembrane proteins
c. Ligand binding opens the interior channel to permit ions to flow into and out of the cell
d. The flow of ions through the ion channel stimulates ligand binding

<u>COMPLETION</u>. Write the word, phrase or number in the blank space in answering the question.

1. Complete the following reactions, indicating the enzyme that catalyzes the reaction and writing structural formulas for products and/or reactants.

a. $(CH_3)_3\overset{+}{N}CH_2CH_2O\overset{\overset{O}{\|}}{C}CH_3 + H_2O \xrightarrow{\text{-------}}$ __________ + __________

b. 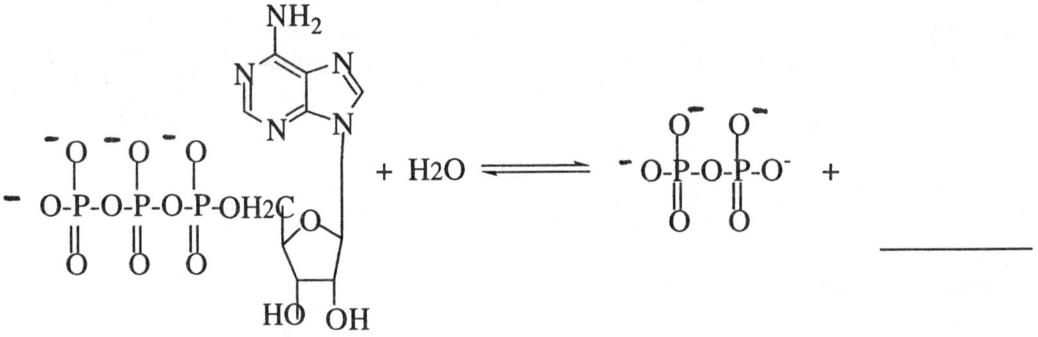

c. __________ + $HOCH_2CH_2\overset{+}{N}(CH_3)_3 \xrightarrow{\text{-------}}$ ________ + ________

2. The __________ is the fluid-filled space between the axon of one neuron and the __________ of the next neuron.

3. A neurotransmitter and a receptor combine very specifically together. This interaction has been likened to the _____ ______ ______ model for enzyme-substrate interactions.

4. Food poisoning from botulism toxin deprives cholinergic nerves of their neurotransmitters. This indicates that the release of __________ from the presynaptic vesicles is blocked.

5. To degrade the neurotransmitter, acetylcholine, in the postsynaptic receptor site, the enzyme, __________, acts on acetylcholine to __________ it to acetic acid and choline.

6. Monoamine neurotransmitters, such as norepinephrine, form a membrane bound _____________ - _________________ like complex, which activates the enzyme __________ that produces the secondary messenger ________________ from ATP.

7. GABA is the abbreviation for ____________________________ .

8. The _________________ end of methionine enkephalin in similar to a segment of the alkaloid painkiller, _________________. Refer to Figure 23.5 in text.

9. Antidiabetic drugs are useful for diabetic patients who have insufficient numbers of ________________ _________________ on their target cells.

10. MAO is the abbreviation for __________, which has the function of _____________ monoamine neurotransmitters by _______ them to an _____________ functional group.

11. The reaction type that converts histidine to histamine is the same type that converts L-Dopa to dopamine. This type of reaction is called a ________ reaction.

12. Match the disease or condition in the left-hand column with the term in the right-hand column that is most closely associated with the disease or condition.

a.	Alzeimer's disease	i.	Produces very low insulin levels
b.	Parkinson's disease	ii.	Tamoxifen binding to estrogen receptors
c.	Impotency	iii.	Low acetylcholine transferase activity
d.	Diabetes Mellitus, type I	iv.	L-Dopa and inhibitors of monoamine oxidase (MAO)
e.	Botulism	v.	Inhibitors of phosphodiesterase activity
f.	Breast cancer	vi.	Prevention of acetylcholine release from presynaptic vesicles

ANSWERS TO SELF-TEST QUESTIONS

Multiple Choice

1. c
2. b
3. c
4. d
5. b
6. c
7. a
8. c
9. a
10. c
11. d
12. b
13. a, b, c

Completion

1. a. acetylcholinesterase; H_3CCO_2H + choline
   b. adenylate cyclase; c-AMP (see figure in text, section 23.5)
   c. $H_3C\overset{\overset{O}{\|}}{C}$-SCoA; acetylcholine transferase; $H_3C\underset{\underset{O}{\|}}{C}OCH_2CH_2\overset{+}{N}(CH_3)_3$ + CoA-SH

2. synapse; dendrites (or cell body)
3. lock-and-key
4. acetylcholine
5. acetylcholinesterase; hydrolyze
6. hormone-receptor; adenylate cyclase; c-AMP
7. γ-aminobutyric acid
8. N-terminal end; morphine
9. insulin receptors
10. monoamine oxidase, inactivating, oxidizing, aldehyde
11. decarboxylation
12. a. iii
    b. iv
    c. v
    d. i
    e. vi
    f. ii

What distinguishes a butterfly from
a lion, a hen from a fly, or a worm
from a whale is........
-F. Jacob
Science, 196, 1161 (1977)

24

Nucelotides, Nucleic Acids, and Heredity

CHAPTER OBJECTIVES

Each one of us is unique in many ways. These specific characteristics result because each individual has a different genetic makeup (genome) in that, although our DNA appears much the same as everyone else's, the sequence (i. e., the order of the bases) of nucleotides in the DNA strands is quite different. These differences become manifest as the genes in the DNA are transcribed into RNA and then translated into the many diverse enzymes and proteins (the real "workers" in the cell), that reveal important aspects of our individuality. In this chapter, we learn about the two types of nucleic acids - DNA and RNA. After first dissecting nucleic acids into their constituents, the general structure of these polynucleotides and the packaging of the DNA in nucleosomes and the higher order structures of chromosomes is examined. The process of replication by which DNA is duplicated with high fidelity is presented, together with critical mechanisms involved with repairing damaged DNA. The process of cloning a segment of DNA is presented, together with a technique called polymerization chain reaction (PCR), which has revolutionized and stream-lined many aspects of cloning. These concepts will provide the basis for examining how selected genes are expressed in cells (Chapter 25) and lead to an understanding of the Central Dogma of Molecular Genetics. The "flow" of genetic information is briefly summarized on the following page and should be consulted during your study of Chapter 25.

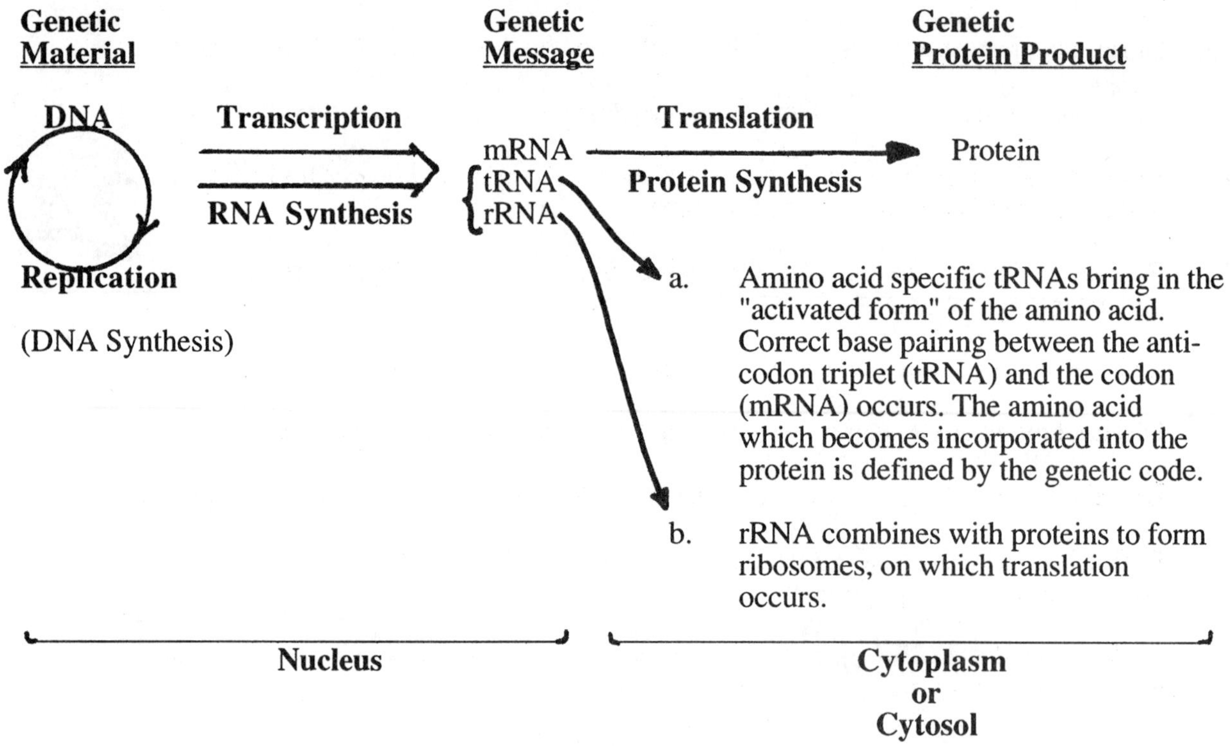

The complexity of DNA is further revealed in the exon and intron segments of the genes in higher organisms and the repeating nucleosome subunits in the nuclear chromosome. These properties were first realized in the mid 1970's.

After you have studied the chapter and worked the assigned exercises in the text and the study guide, you should be able to do the following.

1. Explain the importance of hereditary information and its storage and expression in DNA.

2. Name and draw the structures of the organic bases (purines and pyrimidines), the nucleosides and the nucleotides that occur in DNA and RNA.

3. Draw a dinucleotide unit of DNA containing one purine and one pyrimidine base; point out the 5'- and 3'-ends, the glycosidic bond between the sugar and base and the phosphodiester backbone in the dinucleotide.

4. Draw the complementary base pairs, G≡C, and A=T, showing all atoms and the hydrogen bonding interactions.

5. In double-stranded DNA, indicate the relative positions (inside or outside on the double helix) of the aromatic bases, the sugar and phosphate units in this structure.

6. List the compositional, structural and functional differences between DNA and RNA.

7. Outline the main features of DNA replication and the replisome, indicating the role or significance of unwinding proteins, DNA polymerase, complementary base pairing, semi-conservative replication, 5'---->3' direction, Okazaki fragments and DNA ligase.

8. Describe the structure of a tRNA molecule, the position of the anticodon loop, the role of the 3'-end and the function of this RNA in the translation process.

9. Discuss the role of telomerase in DNA replication and its proposed role in immortality.

10. Define the terms gene, exons and introns as they relate to higher organisms.

11. Describe the composition and structure of the nucleosome and how it folds to results in higher order structure for the chromosome.

12. Explain the immediate goal of the Human Genome Project and its importance in the understanding of human biology.

13. Indicate the important new use of DNA fingerprinting in forensic science.

14. Discuss how pharmacogenomics will help tailor the prescription of drugs.

15. Discuss how DNA repair processes operate to eliminate DNA damage.

16. Define how apoptosis differs from necrosis and indicate the role of capsases and endonuclease in the apoptotic process.

17. Outline how the PCR technique can amplify DNA and indicate the role of temperature, DNA polymerase and primers play in the procedure.

18. Define the important terms and comparisons in this chapter and give specific examples where appropriate.

IMPORTANT TERMS AND COMPARISONS

Chromosomes and Genes
Nucleic Acids
Ribonucleic Acid (RNA)
Nucleus (of the cell)
Pyrimidine Bases; C, T, U
D-Ribose and D-Deoxyribose
Bases, Nucleosides and Nucleotides
Ester and Anhydride Bonds
3'-and 5'-OH Ends of Nucleotides
Anticancer Drugs and Chemotherapy
Complementary Base Pairs
Histone Proteins and Nucleosomes
One Gene-One Enzyme Hypothesis
mRNA, rRNA, tRNA and Ribozymes
Splicing of mRNA
Minisatellite and Microsatellite DNA
Replication Fork
Leading and Lagging DNA Strands
Continuous and Discontinuous Strand Synthesis
Telomeres and Telomerase
DNA Polymerase
Histone Acetylase and Chromatin
Pharmacogenomics and EM, DM and UEM

Heredity
Deoxyribonucleic Acid (DNA)
Transcription and Translation
Cytoplasm
Purine Bases; A, G
β-N-Glycosidic Bond
C-3- and C-5-Carbon on Sugar Ring
Ribonucleotides and Deoxyribonucleotides
A, AMP, ADP and ATP
Primary and Secondary Structure of DNA
A=T and G≡C Hydrogen Bonding
Solenoids and Higher Order Structure
(Inheritable) Gene
Clover Leaf Structure of tRNA
Genes, Exons and Introns
Semiconservative DNA Replication
3'-and 5'-OH Ends of DNA
Okazaki Fragments
Replisomes
Somatic Cells and Germ Cells
Okazaki Fragments
DNA Fingerprinting and Electrophoresis
Human Genome and 30,000 Genes

CYP2D6 Gene and Cytochrome P-450 Enzyme
AP Sites in DNA
Cloning
Hybridization of DNA Strands
DNA Repair, BER and NER
Apoptosis and Necrosis
Polymerase Chain Reaction (PCR)
Restriction Enzymes

Components of the Replisome

Topoisomerase or Gyrases
Helicase
Primase
DNA Polymerase
DNA Ligase

SELF-TEST QUESTIONS

MULTIPLE CHOICE. In the following exercises, select the correct answer from the choices listed. In some cases, two or more choices will be correct.

1. The backbone in a nucleic acid is called a

   a. β–glycosidic bond
   b. sugar backbone
   c. phosphodiester backbone
   d. peptide backbone

2. The coding sequences of a gene in DNA are called

   a. nucleosomes
   b. introns
   c. exons
   d. initiating factors

3. A linkage connecting the base and the deoxyribose is a

   a. peptide backbone
   b. β-N-glycosidic bond
   c. phosphodiester backbone
   d. other linkage

4. A nucleosome is made up of

   a. DNA and histone proteins
   b. rRNA and proteins
   c. nucleosides
   d. both DNA and RNA

5. Telomerase is an enzyme that

   a. is made of both an RNA and protein subunit
   b. replicates the telomeres at the ends of the chromosomes
   c. replicates DNA in a semiconservative manner
   d. can be used in PCR

6. A feature which distinguishes ribose from deoxyribose is the absence of an OH at the

   a. C-2-carbon
   b. C-3-carbon
   c. C-1-carbon
   d. C-5-carbon

7. The hydrogen bonding between a G and C in forming a GC base pair in DNA involves

   a. atoms on or in the six-membered rings of both guanine and cytosine bases
   b. atoms on or in the six-membered-ring of cytosine and the five-membered ring in guanine
   c. only oxygen atoms in cytosine and nitrogen atoms in guanine
   d. only hydrogen-bond donors in cytosine and only hydrogen-bond acceptors in guanine

8. Adenine is the name of a

   a. nucleoside
   b. purine base
   c. nucleotide
   d. amino acid

9. The sequence GCCCTGA has the A at the

   a. 5'-end
   b. 3'-end
   c. can be either the 5'- or 3'- end
   d. depends on size of DNA

10. The 3'-end of a tRNA molecule

   a. contains the anticodon loop
   b. binds to a specific amino acid
   c. has no known function
   d. binds to rRNA

11. The structure of cytosine is

a.

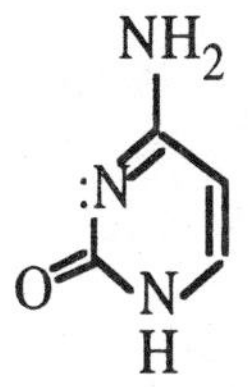

c.

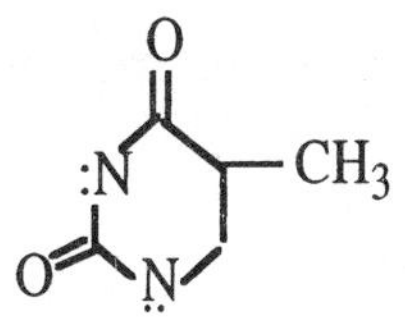

b.

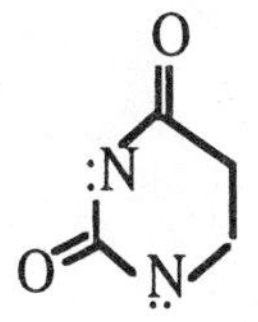

d.

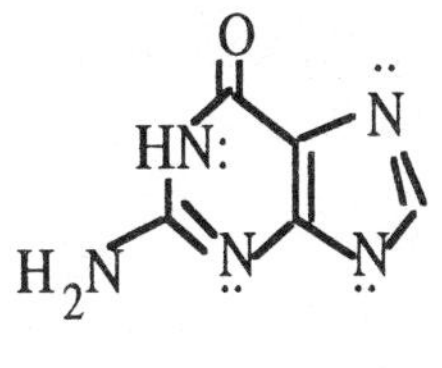

12. DNA fingerprinting requires the use of

   a. protein synthesis
   b. transcription factors
   c. restriction enzymes
   d. exons and introns

13. Okazaki fragments are

a. formed in RNA transcription
b. formed in DNA replication
c. small RNA molecules
d. small DNA fragments of about 1000 nucleotides each

14. A glycosidic bond in a nucleoside or nucleotide links the (deoxy)ribose sugar and which position on a purine base?

a. N-1 atom
b. N-9 atom
c. N-3 atom
d. C-8 atom

15. One of the complementary base pairs in DNA is

a. AU
b. GC
c. AT
d. GT

16. An RNA molecule has no

a. uracil bases
b. thymine bases
c. cytosine bases
d. guanine bases

17. The hydrogen bonding in an AT base pair involves atoms in the purine ring exclusively associated with the

a. 6-membered ring
b. 5-membered ring
c. both the 5- and 6-membered ring
d. none of the above

18. The process by which cells become specialized (neurons, muscle, etc.) in the development of a multicellular organism is called

a. replication
b. differentiation
c. cell division
d. genetic mutation

19. The single-stranded DNA primers used in PCR can be complementary to

a. only the 3-ends of the DNA to be amplified
b. only the 5'-end of the DNA to be amplified
c. either the 3- or the 5-ends of the DNA to be amplified
d. both the 3'-or the 5-ends of the DNA

20 Caspases are

a. nucleases that cut at cytidine residues in DNA
b. enzymes used in cloning of DNA
c. enzymes that cleave proteins next to an aspartic acid residue
d. involved in programmed cell death called apoptosis

21. An RNA nucleoside does not contain a

a. heterocyclic base
b. phosphate
c. ribose sugar
d. amino acid

22. A nucleosome is composed of

a. DNA polymerase and other proteins and enzymes
b. a sugar, heterocyclic base and phosphate
c. 147 base pairs of DNA and eight histone proteins
d. RNA and proteins

23. It is thought that chromatin structure must be opened up before replication can occur. As part of this process,

a. histone deacetylase cleaves off an acetyl group from a serine residue in a histone protein
b. histone acetylase puts an acetyl group onto a lysine reside in a histone protein
c. a phosphate group is put on a histidine residue in a histone protein
d. the histone proteins become glycosylated with a number of sugar groups

COMPLETION. Write the word, phrase or number in the blank space in answering the question.

1. The DNA molecule in humans (the human genome) is thought to have at least _____________ genes.

2. Replication of DNA in eukaryotic cells occurs in the _________of the cell, while translation occurs in the ____________.

3. There are __________ hydrogen bonds between adenine and its (structurally) complementary base, ___________.

4. A nucleic acid that exhibits enzymatic activity is called a ___________ .

5. An RNA-protein complex that exhibits enzymatic activity is called _________ and is associated with cleaving about 50-100 nucleotides from the ________ of chromosomes at each cell division.

6. Amino acids are attached to the _______ end of tRNA on the ____________ unit.

7. Replication of the new DNA strand occurs in a ________________ direction, while the DNA polymerase is traveling in the ________________ direction on the DNA template strand.

8. _____________ are composed of rRNA molecules and proteins and are the units on which protein synthesis occurs in the __________ of the cell.

9. The organic bases in DNA reside on the interior of the double helix and stabilize the structure by _____________ interactions.

10. _____________ bases are made up of two fused rings, while __________ bases have only a six-membered heterocyclic ring.

11. The sequence of one strand of double stranded DNA is 5'-AGGAGCTTCG-3'. The sequence of the complementary strand is ________________________.

12. In BER (repair process), which is an acronym for _______ ______ ______, the enzyme that recognizes DNA damage and cleaves off a base is called ______ ________.

13. Match the scientist(s) in the left-hand column with the accomplishment on the right with which he (they) is most closely associated.

a.	Okazaki	(i)	Proposed the structure of DNA
b.	J. Watson & F. Crick	(ii)	Proposed DNA contains the genetic information
c.	K. Mullis	(iii)	X-ray studies of DNA
d.	O. Avery	(iv)	Proposed one gene-one enzyme hypothesis
e.	R. Franklin & M. Wilkins	(v)	Discovered that in DNA, # of A = of T # of G = # of C
f.	G. Beadle & E. Tatum	(vi)	Discovered PCR technique
g.	E. Chargaff	(vii)	Discovered small segments of DNA made in DNA replication

ANSWERS TO SELF-TEST QUESTIONS

Multiple Choice

1. c
2. c
3. b
4. a
5. a
6. a
7. a
8. b
9. b
10. b
11. a
12. c
13. b
14. b
15. b, c
16. b
17. a
18. b
19. a
20. c, d
21. b
22. c
23. b

Completion

1. 30,000
2. nucleus, cytoplasm (or cytosol)
3. two, thymine
4. ribozyme (or catalytic RNA)
5. telomerase, teleomers
6. 3', adenosine
7. 5'-->3', 3'-->5'
8. Ribosomes, cytoplasm
9. hydrophobic
10. Purine, pyrimidine
11. 3'-TCCTCGAAGC-5'
12. base excision repair, DNA glycosylase
13. a. (vii)
    b. (i)
    c. (vi)
    d. (ii)
    e. (iii)
    f. (iv)
    g. (v)

It is the amazement of selective gene expression
that destines certain cells to become liver cells,
others become heart cells, etc.- all remarkably orchestrated
to make you, me and our neighbor.
-Just thoughts

25

Gene Expression and Protein Synthesis

Each one of us is unique. However, the liver in every one of us functions in the same way, likewise the heart, the colon, etc. These characteristics result from the finding that although every cell contains the same genetic material, DNA, with its many genes, only a small select collection of these genes is expressed into proteins in any particular cell type (tissue). This chapter permits us to begin to understand these remarkable happenings within us.

CHAPTER OBJECTIVES

After you have studied the chapter and worked the assigned exercises in the text and the study guide, you should be able to do the following.

1. Consider a gene on DNA and go through the details of the step-wise processes of transcription and translation (gene -> protein).
2. Outline the transcription process and the role of transcription factors, mRNA and the codon units in the transfer of genetic information.
3. Describe the structure of a tRNA molecule, the position of the anticodon loop, the role of the 3'-end and the function of this RNA in the translation process.
4. Describe the location and role of rRNA in the translation process.
5. Define the genetic code and distinguish between the initiation codon, stop codons and degenerate codons.

6. Describe the processes of amino acid activation and then the initiation, elongation and termination steps in protein synthesis.

7. Describe the role of the A and P sites and the 40S and 60S ribosomes in translation.

8. Define the terms, gene, exons and introns as they relate to higher organisms.

9. Explain how a mutation can be produced during DNA replication or by chemicals called mutagens.

10. Explain the use of plasmids, restriction endonucleases and Escherichia coli in recombinant DNA technology.

11. Describe the role of nucleoside-analog-inhibitors and protease inhibitors in the treatment of patients with AIDS.

12. Describe the role of the promoter, regulatory sequences and regulator transcription factors in the regulation of the expression of eukaryotic structural genes.

13. Indicate the function of chaperone proteins and the proteasome in post-translational controls.

14. Explain the difference between a proto-oncogene and an oncogene and indicate its importance in the life of the cell.

15. Define the important terms and comparisons in this chapter and give specific examples where appropriate.

IMPORTANT TERMS AND COMPARISONS

Central Dogma of Molecular Biology
Gene Expression, Transcription and Translation
Ribonucleic Acid (RNA)
One Gene-One Enzyme Hypothesis
Nucleosome Binding Proteins
(Gene) Promoters
Transcription Factors
Exons and Introns
"Antisense" RNA
Svedberg Unit, S
Codon (Triplet Sequence) in mRNA
Initiation and Termination Codons
Clover Leaf Structure for tRNA
Activated Amino Acid Synthase
40 S and 60 S Ribosomes
Translation: Activation, Initiation, Elongation and Termination
Sieving Portions of AARS
Translocation and Termination
Antibiotics and Antiviral Agents
Nucleoside-Analog Viral Inhibitors
Non-Nucleoside Viral Inhibitors
HIV Glycoproteins - Gp120 and Gp41

Nucleus (of the Cell)
RNA Polymerase I, II and III
mRNA, rRNA and tRNA
Initiation and Termination Signals
Structural and Regulatory Genes
Consensus Sequences and Initiation Signals
Termination Sequence
mRNA Transcript and Mature mRNA
Post-transcriptional Processes
(Almost) Universal Genetic Code
Anticodon (Triplet Sequence) in tRNA
Degenerate or Multiple Codes
Activated Amino Acid
Amino Acid-tRNA
tRNAimet
P and A(Acceptor) Sites on the Ribosomes
Peptidyl Transferease
Stop Codons
DNA Virus and RNA Virus
Retrovirus (RNA) and the AIDS Virus
AZT, ddC, ddI and 3TC
Protease Inhibitors for AIDS Treatment
Vaccine

Translation: Activation, Initiation, Elongation and Termination
Promoter and Operator Sites
TATA Box and GC Box in Promoters
Response Sequences in Promoters
Metal-Binding Fingers, Helix-Turn-Helix And Leucine Zipper Transcription Factors
Chaperones
Nucleotide Excision Repair (NER)
Tumor Suppressor Gene and p53
Recombinant DNA
Plasmids
"Sticky Ends"
Gene Therapy
(Inheritable) Gene
Gene Regulation
Regulatory Sequences and Structural Genes
Transcription Regulatory Factors
Enhancers and Response Elements
Mutagens and Mutations
Mutagens and Carcinogens
Proteasomes
Proto-oncogenes and Oncogenes
Inborn Genetic Disease
Escherichia coli (E. coli) Bacterium
Restriction Endonuclease
DNA Ligase
Genetic Engineering

SELF-TEST QUESTIONS

MULTIPLE CHOICE. In the following exercises, select the correct answer from the choices listed. In some cases, two or more choices will be correct.

1. The central dogma of molecular biology states that

   a. DNA is a double stranded helix
   b. Information flows from DNA to RNA to proteins
   c. Gene expression is different in all tissues
   d. DNA replication is the key event in cell division

2. The coding sequences of a gene in DNA are called

   a. nucleosomes
   b. introns
   c. exons
   d. initiating factors

3. An RNA virus contains which of the following

   a. DNA
   b. RNA
   c. proteins
   d. carbohydrates

4. The RNA molecules that are translated into proteins are

   a. tRNA
   b. rRNA
   c. mRNA
   d. ribozymes

5. The three-letter nucleotide sequence that is listed for the genetic code is the sequence that is found on

   a. codon sequence of mRNA
   b. sequence on DNA
   c. anti-codon sequence of tRNA
   d. codon sequence on rRNA

6. In protein synthesis, the initial amino acid is always

   a. glycine
   b. cysteine
   c. methionine
   d. alanine

7. Translation occurs

   a. on the ribosomes in the cytoplasm
   b. with the first amino acid being the N-terminal residues
   c. on the ribosomes in the nucleus
   d, with the first amino acid being the C-terminal residue

8. The activation of the amino acid prior to translation requires the consumption of how many high-energy phosphate bonds?

   a. none
   b. one
   c. two
   d. three

9. The 3'-end of a tRNA molecule

   a. contains the anticodon loop
   b. binds to a specific amino acid
   c. has no known function
   d. binds to rRNA

10. The bond that links the amino acid residue to AMP is an

   a. ether linkage
   b. acid anhydride linkage
   c. ester linkage
   d. amide linkage

11. The initiating codon in protein synthesis is

   a. GGG
   b. GUA
   c. AUG
   d. ATG

12. Antibiotics are used to

   a. kill viruses
   b. inhibit DNA replication
   c. bind to mutagens
   d. kill bacteria

13. The ribosome unit(s) that contains the P and A sites is (are)

   a. the 60S ribosome
   b. the 40S ribosome
   c. both the 40S and 60 S
   d. on the tRNA

14. The codons that code for the incorporation of the amino acid, cysteine, into a protein is (are)

   a. UUU
   b. UAG
   c. UGU
   d. UGC

15. The interactions in transcription often result in DNA forming a loop of DNA. In these cases, indicate the types of proteins (far from the promoter) that often help in the DNA looping by interacting with protein factors in the promoter.

   a. enhancer binding proteins
   b. RNA polymerase
   c. termination factors
   d. peptidyl transferase

16. In protein synthesis, the mRNA initially attaches to the

a. 30 S ribosome
b. 40 S ribosome
c. 100 S nucleosome
d. 60 S ribosome

17. An mRNA molecule contains no

a. uracil bases
b. thymine bases
c. cytosine bases
d. guanine bases

18. Transcription factors bind to the

a. promoter site
b. structural gene
c. the A site
d. ribosome binding site

19. An amino acid is activated by reaction with

a. tRNA
b. synthetase
c. ATP
d. methionine

20. Restriction endonucleases have the property that they

a. can ligate DNA molecules
b. are required for termination of transcription
c. cleave DNA in a sequence-specific manner
d. bind to regulatory sites in promoters

21. Proteins that contain metal binding sites often also bind to

a. Zn^{+2}
b. Na^{+}
c. H^{+}
d. Fe^{+2}

22. HIV patients are often treated with a three-drug cocktail that contains

a. protease activator
b. protease inhibitors
c. nucleoside analogs
d. transcriptional inhibitors

23. A tumor suppressor gene that is mutated in about 40% of all human cancers is

a. EGF
b. p53
c. p100
d. RNA polymerase

24. A repair mechanism that repairs some mutations is called

a. nuclear repair machinery
b. DNA repairase
c. nucleotide excision repair
d. mutationase

25. A number of transcription factors have been discovered that are identified by the presence of the following features in these proteins:

a. leucine zipper
b. helix-turn-helix motifs
c. metal-binding-fingers
d. all of these features

COMPLETION. Write the word, phrase or number in the blank space in answering the question.

1. The ____ site on the ribosome is the site at which the incoming tRNA binds, while the ______ site is the site where the tRNA binds that contains the growing peptide chain.

2. A mutation by a chemical or by radiation occurs directly on _______, but the result is then transmitted to both the intermediate ________ and the final _________ product.

3. Plasmids are small, ________ DNAs found in ___________ cells. They are useful in the _________ of genes.

4. Human insulin is one of the first proteins to be produced by _____________ _____________ _______________ (three words).

5. It has been found that all __________ are mutatgenic, but not all mutagens are __________.

6. All amino acids are attached to the ____________ end of tRNA on the _________________ unit.

7. Referring to the table for the Genetic Code in the text, write the amino acid sequence for the protein produced from the mRNA shown below: Indicate the N- and C-terminal ends of the protein.

   AUGUUGCACAACGGGGCGGUGGUGUAA

8. After combining the DNA of a plasmid that has sticky ends with another fragment of DNA with the same sticky ends, the enzyme __________________ ____________________ can covalently splice the two DNA fragments together.

9. Ionizing radiation, gamma rays and chemicals, including benzene and vinyl chloride, are known to be __________________.

10. The sequence of one strand of double stranded DNA is 5'-AGGAGCTTCG-3'. The sequence of the RNA strand produced from transcription on this strand is ________________________.

11. Synthesis of RNA (transcription) occurs in a ________________ direction as RNA polymerase moves along the complementary DNA strand in a _____________ direction.

12. The development of cancer may occur when a normal gene, called a ____________, is altered or ___________. The resulting mutated gene is referred to as an ____________.

13. Match the scientist(s) or term in the left-hand column with the most closely associated term in the right-hand column.

a. AIDS	(i) UAA
b. A stop codon	(ii) Determined the genetic code
c. AIDS treatment	(iii) Caused by a retrovirus infection
d. p53	(iv) Cuts DNA at a very specific sequence
e. Restriction enzyme	(v) Proposed the one gene-one enzyme hypothesis
f. G. Beadle & E. Tatum	(vi) Most commonly mutated gene in human cancers
g. M. Nirenberg	(vii) AZT

ANSWERS TO SELF-TEST QUESTIONS

Multiple Choice

1. b
2. c
3. b, c
4. c
5. a
6. c
7. a, b
8. c
9. b
10. b
11. c
12. d
13. c
14. c, d
15. a
16. b
17. b
18. a
19. c
20. c
21. a
22. b, c
23. b
24. c
25. d

Completion

1. A, P
2. DNA, RNA, protein
3. circular, bacterial (E. coli), cloning
4. recombinant DNA technology
5. carcinogens, carcinogenic
6. 3', adenosine
7. (N-terminal) Met-Leu-His-Asn-Gly-Ala-Val-Val (C-terminal)
8. DNA ligase
9. mutagenic and carcinogenic
10. 3'-TCCTCGAAGC-5'
11. 5'------>3', 3'------>5'
12. proto-oncogene, mutated, oncogene
13. a. iii
    b. i
    c. vii
    d. vi
    e. iv
    f. v
    g. ii

"Whatever a cell does has to be paid
for in the currency of energy".
- Szent-Gyorgyi

26

Bioenergetics: How the Body Converts Food to Energy

CHAPTER OBJECTIVES

After you have studied the chapter and worked the assigned exercises in the text and the study guide, you should be able to do the following.

1. List the major organelles in a typical animal cell and indicate at least one important biological process that is carried out in each organelle.

2. List the (molecular) components contained in ATP, the coenzymes, NAD^+ and FAD, and acetyl CoA. Write out a general reaction in which each takes part and draw the part of each structure that is changed as a result of the reaction.

3. Distinguish generally between the catabolic and the anabolic process in terms of the consumption or production of ATP, NAD^+ and FAD.

4. Draw a clear schematic representation of a mitochondrion pointing out the outer membrane, inner membrane, intermembrane region, the matrix, cristae, and the location of the proton translocating ATP-ase.

5. State the major role of ATP in metabolic reactions.

6. Explain the role of NAD^+ and FAD in the citric acid cycle and in the electron transport chain.

7. Describe the role of acetyl CoA in the Krebs cycle.

8. Name the major species in the Krebs cycle.

9. Indicate the type of reaction involved at each step in the Krebs cycle and indicate (in words) an example of each.

10. Explain how the reduced coenzymes, NADH and $FADH_2$, are involved in oxidative phosphorylation and why each produces a different number of ATP molecules.

11. Determine the number of ATP molecules produced in the Krebs cycle and oxidative phosphorylation as a result of the oxidation of one acetyl group.

12. Describe the important features of the chemiosmotic hypothesis.

13. Describe the function of the proton translocating ATP-ase.

14. Explain the value of using an uncoupling agent in oxidative phosphorylation studies.

15. Give specific examples of the use of ATP to produce mechanical, electrical and chemical energy.

16. Describe the roles of the enzymes, cytochrome P-450, superoxide dismutase and catalase in specific oxidative reactions.

17. Define the important terms and comparisons in this chapter and give specific examples where appropriate.

IMPORTANT TERMS AND COMPARISONS

Metabolism: Catabolism and Anabolism
Common Catabolic Pathway
Organelles
Riboflavin and Flavin
Coenzymes
Mitochondrion-Outer and Inner Membrane, Intermembrane Space, Matrix and Cristae
Activated C_2 Fragments or Acetyl Coenzyme A
Turn Over Rate for ATP
High-Energy Phosphate Anhydride Bond
Nicotinamide and Riboflavin
Mercaptoethylamine
Dehydration Reaction
Tertiary and Secondary Alcohol
GTP and ATP: Energetically Equivalent
Coenzyme Q
Uncoupling Agents
Proton Translocating ATP-ase
Cytochrome P-450
Superoxide (Anion), O_2^-
Catalase
Energy Yield
Myosin and Actin

Metabolic or Biochemical Pathways
Energy Carrier Molecule, ATP
Nucleus, Lysosomes, Golgi Bodies, Mitochondria
Flavoproteins
NADH and NAD^+ Couple
$FADH_2$ and FAD Couple
Coenzyme A, Acetyl Group (CH_3CO-) and Acetyl Coenzyme A
AMP, ADP and ATP
Pantothenic Acid
Krebs Cycle, Tricarboxylic Acid Cycle and Citric Acid Cycle
Hans Krebs and Peter Mitchell
Oxidative Decarboxylation
Cytochrome c
Chemiosmotic Theory
Proton Gradient and Proton Channel
Protonphore
Superoxide Dismutase (SOD)
Glutathione
Energy Conversions
$1/2\ O_2 + 2\ H^+ + 2\ e^- = H_2O$

FOCUSED REVIEW

I. Stages of Metabolism

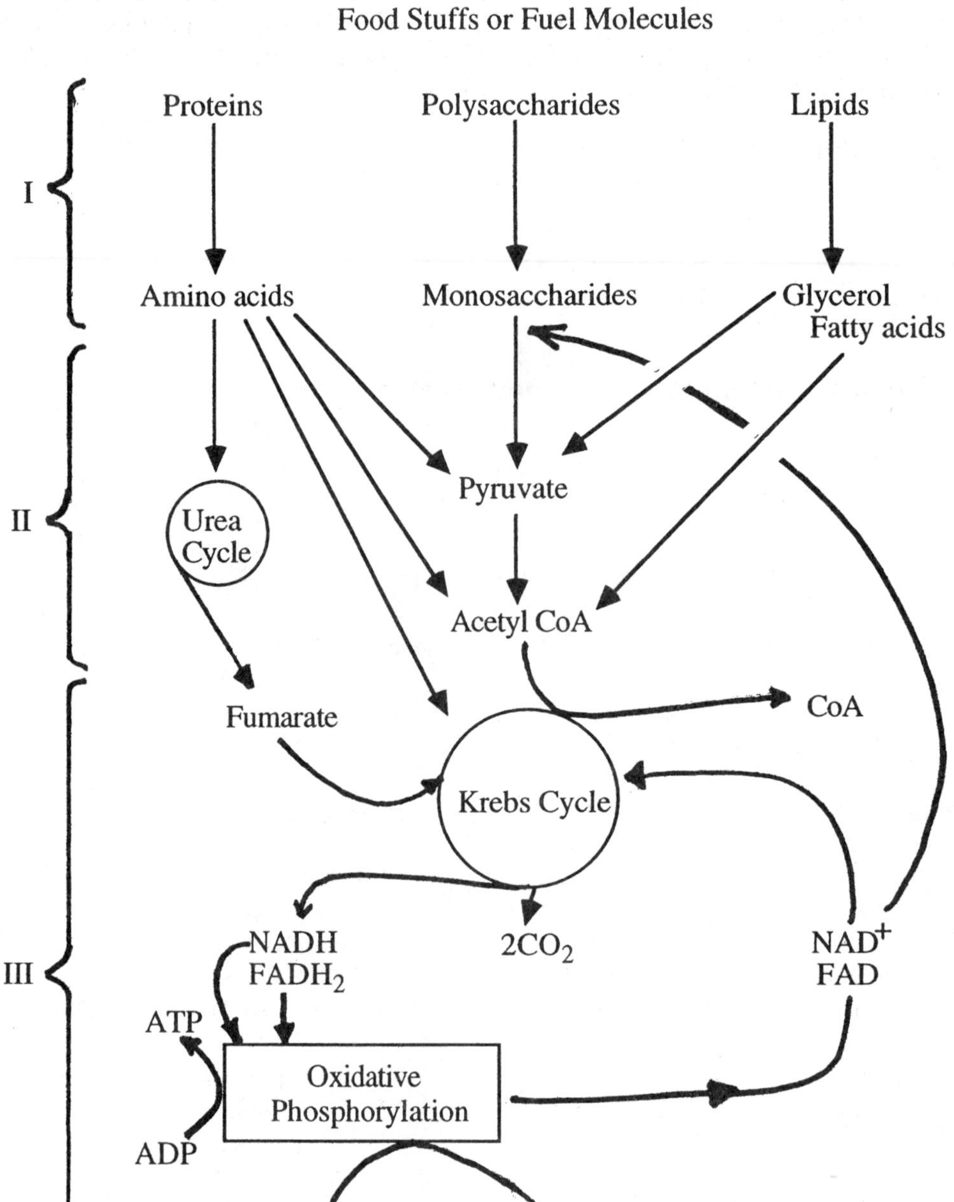

In the initial stage, the macromolecules in food are digested to monomers. Proteins are degraded to amino acids, polysaccharides to glucose and other monosaccharides and lipids are hydrolyzed to fatty acids and glycerol. In the second stage, the sugars, fatty acids, glycerol and many of the amino acids are converted into the activated acetyl units (C_2 fragments) of acetyl coenzyme A. The nitrogen atoms in amino acids are processed in the urea cycle, while the deaminated carbon skeletons are processed directly or in a few steps into the Krebs cycle. The major focus in stage III involves the Krebs cycle and oxidative phosphorylation. In these final stages of the oxidation of "fuel" molecules, the C_2 fragment of acetyl CoA enters the Krebs cycle and is completely oxidized to 2 CO_2, reduced coenzymes NADH and $FADH_2$ are produced, along with one GTP molecule.

The reduced coenzymes, NADH and $FADH_2$, enter in the electron transport chain and are oxidized to NAD^+ and FAD, which are then available for recycling into the Krebs cycle and other catabolic pathways, including the glycolysis and the β-oxidation of fatty acids pathways (Chapter 27). As a result of the oxidation, hydrogen ions are expelled from the mitochondria and upon re-entry produce ATP from the phosphorylation of ADP. The electrons and the hydrogen ions now combine with O_2 to form water. Note that most of the ATP generated from the catabolic metabolism of the foodstuffs is formed in stage III.

II. Summary of Reactions

a. Interconversion of ATP to ADP + Pi

$$\text{ATP} \rightleftharpoons \text{ADP} + \text{Pi}$$

b. Half-reactions for coenzymes NAD^+ and FAD

$$NAD^+ + H^+ + 2\,e^- = NADH$$
$$FAD + 2\,H^+ + 2\,e^- = FADH_2$$

c. Overall reaction in Krebs cycle

$$H_3CCOOH + 2\,H_2O + 3\,NAD^+ + FAD + GDP + Pi \longrightarrow$$

$$2\,CO_2 + 3\,NADH + FADH_2 + GTP + 3\,H^+$$

a waste product	to electron transport chain	energy

d. Overall reaction of oxidative phosphorylation

$$NADH + 3\,ADP + 1/2\,O_2 + 3\,Pi + H^+ = NAD^+ + 3\,ATP + 4\,H_2O$$
$$FADH_2 + 2\,ADP + 1/2\,O_2 + 2\,Pi = FAD + 2\,ATP + 3\,H_2O$$

from Krebs cycle	recycle back to Krebs cycle	energy

e. Oxidation of a C_2 (acetyl) fragment

$$C_2 + 2\,O_2 + 12\,ADP + 12\,Pi = 12\,ATP + 2\,CO_2$$

III. Citric Acid Cycle*

Step	Reaction	Type
1.	Acetyl CoA + oxaloacetate + H_2O -----> citrate + CoA oxaloacetate: $(\underline{C}_4\underline{H}_2\underline{O}_5)$; citrate: $(\underline{C}_6\underline{H}_5\underline{O}_7)$	Condensation reaction
2.	Citrate -----> cis-aconitate + H_2O -----> Isocitrate Citrate: $(C_6H_5O_7)$ tertiary alcohol; Isocitrate: $(C_6H_5O_7)$ secondary alcohol	Isomerization, by way of a dehydration, followed by a hydration
3.	Isocitrate + NAD^+ -----> α-ketoglutarate + NADH + CO_2 Isocitrate: $(\underline{C}_6\underline{H}_5\underline{O}_7)$ alcohol; α-ketoglutarate: $(\underline{C}_5\underline{H}_4\underline{O}_5)$ ketone	Oxidative decarboxylation
4/5	α-ketoglutarate + NAD^+ + H_2O + GDP + Pi -----> succinate + CO_2 + NADH + H^+ + GTP α-ketoglutarate: $(\underline{C}_5H_4\underline{O}_5)$ ketone; succinate: $(\underline{C}_4H_4\underline{O}_4)$ acid	Oxidative decarboxylation
6.	Succinate + FAD -----> Fumarate + $FADH_2$ Succinate: $(C_4\underline{H}_4O_4)$; Fumarate: $(C_4\underline{H}_2O_5)$	Oxidation (of succinate)
7.	Fumarate + H_2O -----> malate Fumarate: $(C_4\underline{H}_2\underline{O}_4)$; malate: $(C_4H_4\underline{O}_5)$	Hydration
8.	Malate + NAD^+ -----> oxaloacetate + NADH + H^+ Malate: $(C_4\underline{H}_4O_5)$ alcohol; oxaloacetate: $(C_4\underline{H}_2O_5)$ ketone	Oxidation (of malate)

*The Krebs cycle can be regarded as a catalytic cycle in which a 2-carbon fragment enters the cycle in the form of acetic acid (an acetyl group plus the OH from H_2O), is oxidized and leaves the cycle as CO_2.

The major species in each step of the cycle can be easily remembered by using the following saying as a mnemonic device. "**O**f **C**ourse, **I** **K**now **S**ome **F**amous **M**en". The first letter in each substrate - **O**xaloacetate, **C**itrate, **I**socitrate, α-**K**etoglutarate, **S**uccinate, **F**umarate and **M**alate begins with these same letters as in the saying.

Again, the overall reaction in the Krebs cycle is:

$$CH_3COOH + 2\ H_2O + 3\ NAD^+ + FAD + GDP + P_i \longrightarrow 2\ CO_2 + 3\ NADH + FADH_2 + 3\ H^+ + GTP$$

IV. Mitochondrion Components

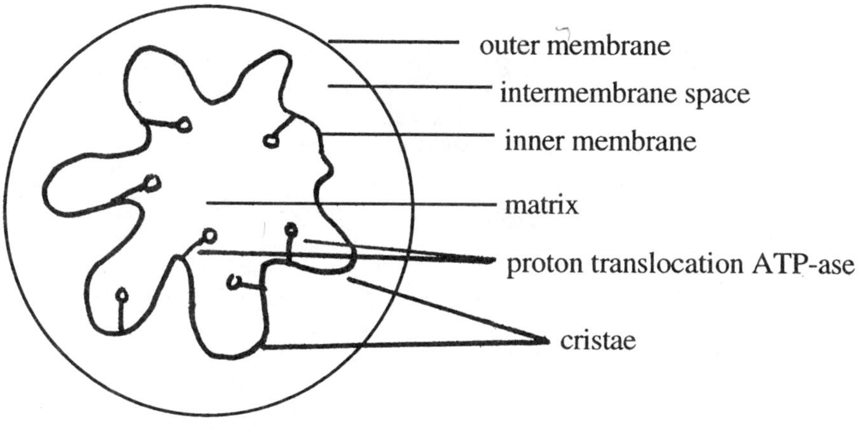

SELF-TEST QUESTIONS

<u>MULTIPLE CHOICE.</u> In the following exercises, select the correct answer from the choices listed. In some cases, two or more answers will be correct.

1. $FADH_2$ transfers its electrons into the electron transport chain in the mitochondrion directly to

   a. ATP-ase
   b. Flavoprotein
   c. Q Enzyme
   d. Cytochrome c

2. Indicate which of the following processes does not consume ATP.

   a. Muscle contraction
   b. Flow of electrons into mitochondria through proton translocating ATP-ase
   c. Catabolism
   d. Pumping of K^+ into the cells by transport proteins

3. In the reaction in which succinate is converted to fumarate in the TCA cycle, succinate undergoes a

a. Hydration
b. Oxidation
c. Isomerization
d. Oxidative decarboxylation

4. Which of the following molecules are produced in the Krebs cycle?

a. NAD^+
b. CO_2
c. GDP
d. Oxaloacetate

5. The proton translocating ATP-ase is located where in the mitochondrion?

a. Intermembrane space
b. Outer membrane
c. Inner membrane/matrix
d. Outside the inner membrane

6. The number of NADH molecules generated in one cycle of the Krebs cycle is

a. 1
b. 2
c. 3
d. 4

7. The amount of energy produced by the hydrolysis of ATP is

a. 7.3 kcal/mole
b. 3.4 kcal/mole
c. 14.6 kcal/mole
d. less than 1 kcal/mole

8. The reduction of FAD to $FADH_2$ occurs in which part of the FAD molecule?

a. Ribose
b. Flavin
c. ADP
d. Nicotinamide

9. Indicate which of the following molecules are carriers of intermediates in metabolism.

a. CoA
b. ADP
c. FAD
d. Nicotinic acid

10. Of the following molecules, which contain high energy bonds?

a. ATP
b. AMP
c. CO_2
d. Isocitrate

11. The number of electrons, originating from the electron transport chain, that is needed to produce one mole of water is:

a. 1
b. 2
c. 3
d. 4

12. Niacin is another name for

a. Nicotinic acid
b. α-ketoglutarate
c. Q enzyme
d. Nicotinamide

13. Which of the following terms is NOT another name for the Krebs cycle?

    a. Tricarboxylic acid cycle
    b. Electron transport chain
    c. Oxidative decarboxylation
    d. Citric acid cycle

14. The acetyl group in acetyl CoA is linked through what kind of a bond?

    a. (C-C) bond
    b. (C-S) bond
    c. (C-O) bond
    d. (C-N) bond

15. Which of the following molecules are NOT an electron-carrier in the mitochondrial membrane?

    a. Cytochrome a_3
    b. Q enzyme
    c. Fumarase
    d. Citrate synthetase

16. For each two C_2 fragment that enters the TCA (tricarboxylic acid) cycle, the total number of ATP and (GTP) molecules produced in the cell is

    a. 24
    b. 22
    c. 10
    d. 12

<u>COMPLETION</u>. Write the word, phrase or number in the blank or draw the appropriate structure in answering the question.

1. A ________________ is a compound that permits H^+ ions to pass through the ________________ membrane passively.

2. The correct order for the enzymes, Q enzymes, FeS protein, cytochromes and flavoprotein in the electron transport chain is

   ______________, ________________, ______________, ________________

3. Anabolism is associated with the ______________ of molecules, while catabolism deals with the ______________ of molecules.

4. ________________ is the molecular unit which is common in FAD, NAD^+, ATP and coenzyme A.

5. For each NADH molecule, a total of ________________ pairs of protons are pumped out of the mitochondrion into the ________________ ________________, with the resultant production of ________________ molecules of ATP.

6. Identify the 6 most important features of the mitochondrion

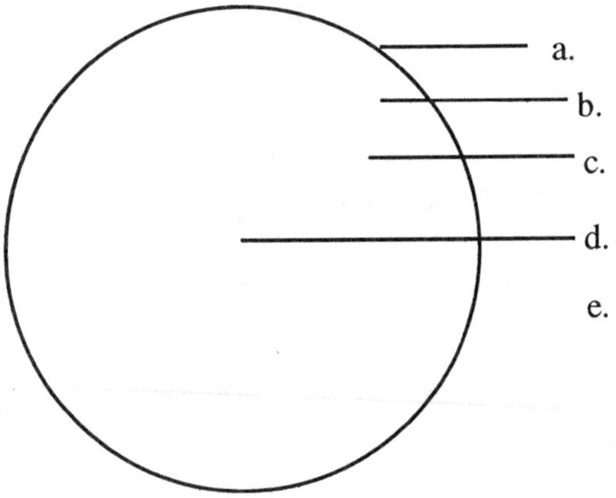

7. Considering the four intermediate carriers, ______________ is a molecular unit unique to FAD, while ______________ is a unique unit occurring in coenzyme A.

8. In the citric acid cycle, the tertiary alcohol in ___________ must be converted to a secondary alcohol in ______________ before the subsequent oxidation to produce a ______________ (functional) group can occur.

9. The ______________ hypothesis, proposed by Peter Mitchell states (in a short paragraph).

10. The reactions of the Krebs cycle occur in the ______________ (an organelle).

11. The products of the reaction of acetyl CoA and oxaloacetate are ______________ and ______________.

12. The reactions in the electron transport chain eventually lead to the reaction of O_2 with ______________ and ______________.

13. NAD^+ is a coenzyme that can act as an ____________ agent in the Krebs cycle. In the process of being converted to the NADH form, it becomes __________ and picks up __________ (a number) proton and ________ (a number) electrons.

14. The major organelles and a primary function of each is:
The _________ is where DNA replication takes place.
The removal of damaged cellular components occurs in the ___________.
__________ ________ are responsible for the packaging and processing of proteins which are to be secreted from the cell.

15. The two critical intracellular enzymes that are responsible for decomposing the highly reactive superoxide anion are _________ _________ and ___________.

ANSWERS TO SELF-TEST QUESTIONS

Multiple Choice

1. c
2. b, c
3. b
4. b, d
5. c
6. c
7. a
8. b
9. a, c
10. a
11. b
12. a
13. b, c
14. b
15. c, d
16. a

Completion

1. protonphore, mitochondrial
2. flavoprotein, FeS protein, Q enzyme, cytochrome
3. synthesis, breaking down
4. ADP
5. three, intermembrane space, three
6. Refer to the figure in Section IV of Focused Review
   a. outer membrane
   b. intermembrane space
   c. inner membrane
   d. matrix
   e. proton translocating ATP-ase
   f. cristae
7. flavin, pantothenic acid
8. citric acid, isocitrate, ketone
9. chemiosmotic, the energy in the electron transfer chain creates a proton gradient. There is a higher concentration of H^+ in the intermembrane space than inside the mitochondrial matrix. The protons pumped out of the matrix provide the driving force for the phosphorylation of ADP by driving the protons back into the mitochondrial matrix through the proton translocating ATP-ase. This enzyme manufactures the ATP.
10. mitochondrion
11. citrate, CoA
12. H^+, electrons
13. oxidizing, reduced, one, two
14. nucleus, lysosomes, Golgi bodies
15. superoxide dismutase (SOD), catalase

"You are what you eat."
- Unknown

27

Specific Catabolic Pathways: Carbohydrate, Lipid, and Protein Metabolism

CHAPTER OBJECTIVES

After you have studied the chapter and worked the assigned exercises in the text and the study guide, you should be able to do the following.

1. Name the major pathways by which energy is extracted from monosaccharides and fatty acids.

2. Discuss the three stages of glycolysis with regard to the energetics (ATP consumption or generation), the number of carbons in the molecular units and the reaction types involved.

3. Distinguish between anaerobic and aerobic glycolysis.

4. Indicate under what conditions pyruvate is converted to (a) lactate, (b) ethanol or (c) acetyl CoA. Write out the reaction for each process.

5. Tabulate the energy yield produced by reactions involved in the complete metabolism of glucose. Indicate the influence of the glycerol-3-phosphate or the aspartate-malate shuttle in the total energy yield.

6. Indicate the point in the catabolic process at which (a) glycerol and (b) fatty acids converge to a common pathway with monosaccharides.

7. Indicate the types of reactions in the β-oxidation spiral of fatty acids.

8. Specify the cellular location in which (a) glycolysis, (b) fatty acid activation and oxidation, (c) citric acid cycle, (d) oxidative phosphorylation and the (e) urea acid cycle occur.

9. Indicate the role of the pentose phosphate pathway in general anabolic metabolism and nucleic acid metabolism, in addition to its indirect role in maintaining red blood cells.

10. Outline the role of phosphorylation in the regulation of fatty acid metabolism.

11. Compare the energy yields for the complete oxidation of (a) glucose, (b) glycerol and (c) a fatty acid.

12. Name the ketone bodies and explain why they are found in high concentrations in the blood under certain conditions.

13. Discuss the three stages of nitrogen metabolism.

14. Indicate the relationship between heme, biliverdin, bilirubin, and urobilin.

15. Define the important terms and comparisons in this chapter and give specific examples where appropriate.

IMPORTANT TERMS AND COMPARISONS

Food
Fat Storage Depots
Common Catabolic Pathway
Anaerobic Pathway
Pentose Phosphate Pathway
Glycerol
Energy Yield for Glucose
β-Oxidation Spiral (of Fatty Acids)
Carnitine Acyltransferase
Stearic Acid
Ketoacidosis
Lipomics
Transamination and Oxidative Deamination
α-Ketoglutarate
Carbamoyl Phosphate
Ubiquitin, Proteasome and Protein Degradation
Phenylketonuria (PKU)
Jaundice and Bilirubin Levels
Glycerol-3-Phosphate Transport (Skeletal Muscles and Nerves)
Aspartate-Malate Shuttle (Heart and Liver)
Carbohydrates, Lipids and Proteins
Amino Acid Pool
Glycolysis
Kinases, Isomerases and Aldolase
Glucose 6-Phosphate Conversion to Ribose and NADPH
Glucogenolysis: Glycogen to Glucose
Mitochondria
Thiokinase
Ketone Bodies: Acetoacetate, Acetone and β-Hydroxybutyrate
Amino Acid Pool
Urea Cycle: Production of Urea From CO_2 and NH_3
Ornithine
Amino Acids: Glucogenic and Ketogenic
Heme Catabolism: Heme, Biliverdine, Bilirubin and Urobilin
Ferritin
Acyl-CoA and Acetyl-CoA

SELF-TEST QUESTIONS

MULTIPLE CHOICE. In the following exercises, select the correct answer from the choices listed. In some cases, two or more answers will be correct.

1. The initial reaction in the β-oxidation spiral of fatty acids produces

a. acyl CoA
b. acetyl CoA
c. carbamoyl phosphate
d. dihydroxyacetone phosphate

2. The final product of aerobic glycolysis is

a. pyruvate
b. acetyl CoA
c. lactate
d. ethanol

3. The number of ATP molecules produced on the complete metabolism of glucose in skeletal muscle is

a. 24
b. 36
c. 12
d. 38

4. Anaerobic glycolysis occurs in the

a. mitochondria
b. cytosol
c. cytosol and mitochondria
d. extracellular fluid

5. Ketone bodies are produced in conditions of a

a. high glucose supply
b. low amino acid pool
c. low glucose supply
d. missing fructose-1-phosphate aldolase

6. The heme unit of hemoglobin contains

a. Fe^{+2}
b. globin protein
c. carbohydrate
d. glycerol

7. The first reaction involved in nitrogen catabolism of amino acids is a(n)

a. oxidative deamination
b. reduction
c. hydration
d. transamination

8. The final step in the urea cycle involves the

a. splitting of argininosuccinate
b. oxidation of glutamate
c. hydrolysis of arginine
d. transamination of glutamate

9. The order of the caloric value for carbohydrates, proteins and fats is

a. protein = fat = carbohydrate
b. carbohydrate > fat > protein
c. fat > carbohydrate > protein
d. protein > fat > carbohydrate

10. Pyruvate undergoes an oxidative decarboxylation with CoA to produce acetyl CoA and

a. lactate
b. glutamate
c. CO_2
d. NAD^+

11. Which of the following molecules is transported into the mitochondria by either the glycerol-3-phosphate transport or the aspartate-malate shuttle?

a. ATP
b. CO_2
c. NADH
d. ADP

12. The number of cycles of the β-oxidation spiral that the (saturated) palmitic acid (C_{16}) goes through is

a. 8
b. 16
c. 7
d. 4

13 A main function(s) of the pentose phosphate pathway is (are)

a. generation of ribose
b. generation of NADPH
c. generation of NADH
d. generation of $NADP^+$

14. Carnitine acyltransferase is essential in the functioning of the

a, pentose phosphate pathway
b. fatty acid metabolism
c. urea cycle
d. ketone body production

COMPLETION. Write the word, phrase or number in the blank space in answering the question.

1. The three end products of complete catabolism of proteins in the body are ________, ________, and ______________.

2. Specialized cells that store fats are called ________________________________.

3. In the catabolic metabolism of hemoglobin, the iron is taken up by ____________.

4. __________________ is considered a ketone body despite not having a ketone functional group.

5. An aldolase enzyme catalyzes the splitting of fructose 1, 6-bisphosphate to _______________ and _________________.

6. Fats are metabolized by first hydrolyzing them to _________ and _______________.

7. Although ____________ is a basic amino acid and occurs in the urea cycle, it is not found in proteins.

8. The β-oxidation of a C_{10} unsaturated fatty acid requires one ____________ step in the process than that for the saturated C_{10} fatty acid.

9. Fatty acid oxidation occurs in the ____________________.

10. If the carbon skeleton of an amino acid is converted to pyruvate, it may serve in either of two roles.

a. ____________________________

b. ____________________________

11. Damaged proteins are covalently modified by reaction with another protein called _______________, after which this "flagged" or modified protein is delivered to the machinery for proteolysis, called the _______________.

12. When the levels of glucose are high, fatty acid oxidation is ___________. On the other hand, , in starvation conditions, fatty acids are ________. However, the ___________ produced from fatty acid oxidation cannot enter the TCA cycle because there is no _____occurring. As a result, _____ bodies are produced and used as a source of _______.

ANSWERS TO SELF-TEST QUESTIONS

Multiple Choice

1. a
2. a
3. b
4. b
5. c
6. a
7. d
8. c
9. c
10. c
11. c
12. c
13. a, b
14. b

Completion

1. urea or NH_4^+, CO_2, H_2O
2. fat deposits
3. ferritin
4. β-hydroxybutyrate
5. glyceraldehyde 3-phosphate, dihydroxyacetone phosphate
6. glycerol, fatty acid
7. ornithine
8. more
9. mitochondrial matrix
10. a. an energy source
    b. a building block in the synthesis of glucose (and other metabolites)
11. ubiquitin, proteasome
12. inhibited, oxidized, acetyl CoA, glycolysis, ketone, energy

If we could not store so vital a substance as glucose in our bodies, we would have to be eating incessently to maintain a steady supply of it. That would be a precarious existence.

-Ernest Borek
The Atoms Within Us

28

Biosynthetic Pathways

CHAPTER OBJECTIVES

After you have studied the chapter and worked the assigned exercises in the text and the study guide, you should be able to do the following.

1. Contrast anabolic and catabolic reactions in terms of energy requirements, the form of coenzymes involved and the cellular location in which they generally take place.

2. Explain why anabolic reaction pathways are usually not just the reverse of the corresponding catabolic pathways.

3. Using Figure 28.1 (text), identify the major metabolic intermediates that may be used in gluconeogenesis and indicate how glycolysis differs from gluconeogenesis.

4. Discuss the importance of UDP-glucose in carbohydrate biosynthesis.

5. Identify the feature that is common in both fatty acid synthesis and degradation.

6. Describe the role of acetyl CoA and malonyl CoA in fatty acid synthesis.

7. Indicate the form of CoA and the other reactants involved in the synthesis of phosphatidate and sphingosine.

8. Write out the reactions for the synthesis of glutamic acid from α-ketoglutaric acid and the subsequent transamination reaction to form other non-essential amino acid.

9. Distinguish the difference between essential and nonessential amino acids.

10. Outline the photosynthetic process in terms of a light reaction and a dark reaction.

11. Define the important terms and comparisons in this chapter and give specific examples where appropriate.

IMPORTANT TERMS AND COMPARISONS

Le Chatelier's Principle
Glycogen
Chloroplasts
Light and Dark Reactions
Glycogenesis and Glycogen
Uridine Diphosphate (UDP)
HMG Reductase and Cholesterol Biosynthesis
Ras Protein, Signal Transduction and Cancer
Essential Fatty Acid
Malonyl ACP
Glucogenic Amino Acid
Ribulose 1, 5-**Bis**phosphate **C**arboxylase -**O**xygenase (**RuBisCO**)
Anabolic and Biosynthetic Pathway
Photosynthesis: Photosystem I and Photosystem II
Calvin Cycle
Acyl Carrier Protein (ACP)
Acyl CoA and Acetyl CoA
Geranyl and Farnesyl Pyrophosphates
Prenylation
Fatty Acid Synthase
Gluconeogenesis and Glycogenesis
Essential and Non-Essential Amino Acids

SELF-TEST QUESTIONS

MULTIPLE CHOICE. In the following exercises, select the correct answer from the choices listed. In some cases, two or more choices will be correct.

1. Gluconeogenesis is associated with the synthesis of

a. glycogen
b. glucogenic amino acids
c. glucose
d. glucose-1-phosphate

2. Glycolysis and gluconeogenesis proceed in reverse directions. However, the number of points in this pathway in which there are enzymes not common to each pathway is

a. 6
b. 8
c. 4
d. 2

3. Both fatty acid synthesis and degradation have the common feature of involving

a. ATP
b. NAD^+
c. acetyl CoA
d. ketogenic amino acids

4. In each cycle in fatty acid synthesis, the unit added on is a

a. C_1 fragment
b. C_2 fragment
c. C_4 fragment
d. C_6 fragment

5. Which of the following structures represents UDP-glucose?

a.

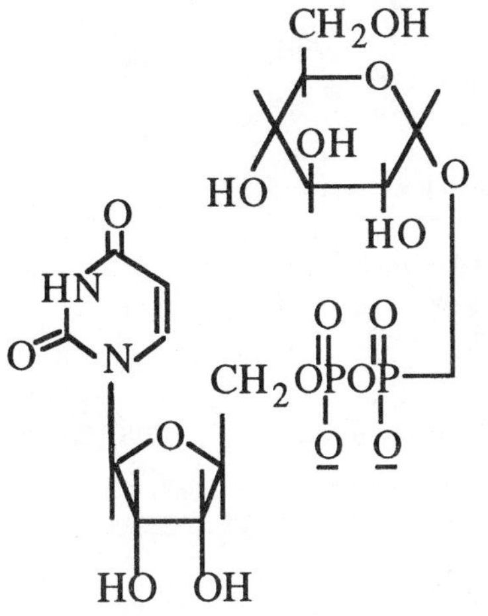

c.

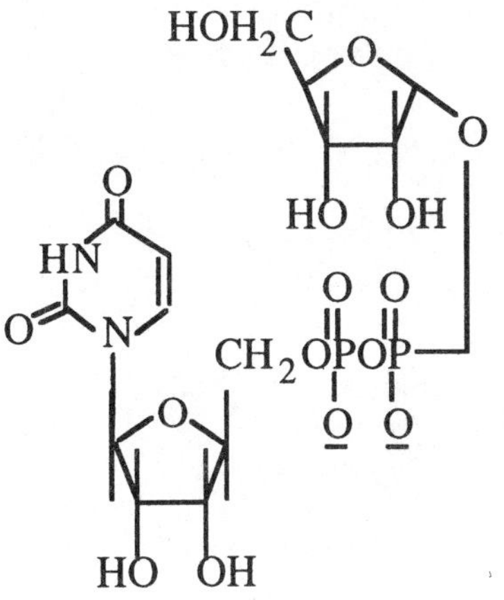

b.

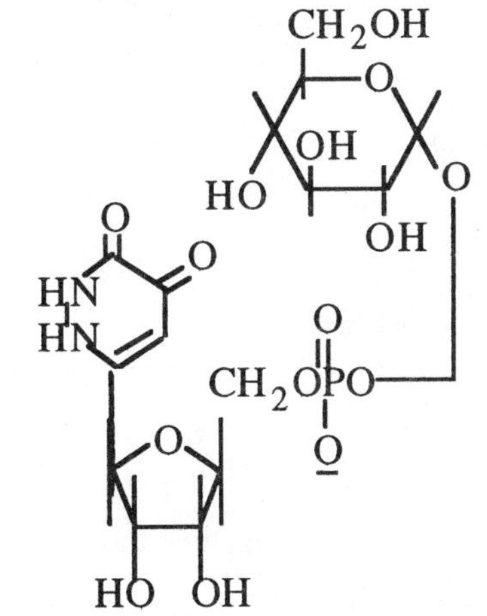

d.

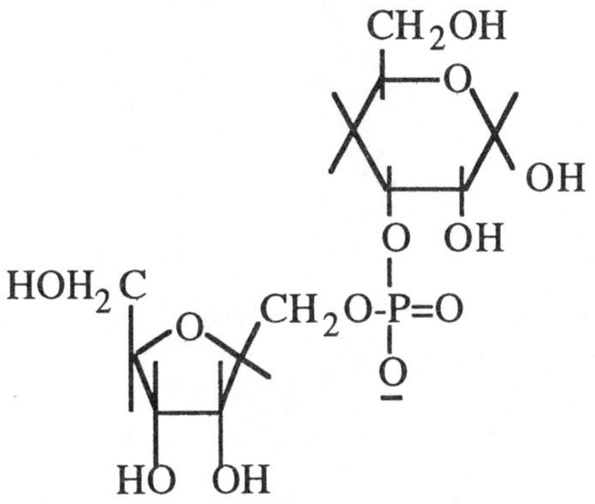

6. The amino acid produced in the following reaction is:

CH_3		CO_2^-	
|		|	
C=O		$HCNH_3^+$	------>
|	+	|	<------
CO_2^-		CH_2	
		|	
		CH_2	
		|	
		CO_2^-	

a. glycine
b. glutamic acid
c. arginine
d. alanine

7. Which of the following can be used in gluconeogenesis?

a. pyruvate
b. lactate
c. ketogenic amino acids
d. urea

8. Which of the following are essential fatty acids?

a. stearic acid
b. butyric acid
c. palmitic acid
d. linoleic acid

9. The Calvin cycle is associated with

a. the dark reaction
b. requires UDP-glucose
c. utilizes RuBisCO
d. converts H2O to O2

COMPLETION. Write the word, phrase or number in the blank space in answering the question.

1. Biosynthetic pathways require an ___________ input and require _____________ power in the form of NADPH.

2. Glycogen is synthesized from the activated glucose monomer, _________________, which is formed in the reaction of __________ and __________________.

3. The overall reaction in photosynthesis is ____________________________________.

4. Fatty acid synthesis occurs in the _____________ , while fatty acid degradation takes place in the _____________ .

5. The acetyl or malonyl group in the ACP complex is attached through a __________ atom.

6. The fatty acid component of phosphatidate is derived from the ____________ which reacts with glycerol-3-phosphate.

7. The biosynthesis of cholesterol occurs in the _____________ (an organ) and is assembled from ______________.

8. The phosphatidylcholine in normal lungs contains predominantly ________________ fatty acids.

9. Photosynthesis occurs in the _________ organelles of plants.
The first reaction in photosynthesis requires energy (sunlight) and is called the _________ reaction. The four important species generated are ________ , ________ , ___________ and __________ .
The second reaction uses CO_2 and three of the species generated in the light reaction , ________ , ___________ and ________ to produce _____________ macromolecules.

10. Most anabolic reactions occur in the ______________.

11. Amino acids that can be synthesized in the human body are referred to as _____________ amino acids.

12. Cholesterol biosynthesis involves three intermediates that contain a C_5, C_{10} and C_{15} unit, respectively, bound directly to pyrophosphate. The names of these three units are _________________, ____________ and ___________.

ANSWERS TO SELF-TEST QUESTIONS

Multiple Choice

1. c
2. c
3. c
4. b
5. a
6. d
7. a, b
8. d
9. a, c

Completion

1. energy, reducing
2. UDP-glucose, UTP, glucose-1-phosphate
3. $n\,H_2O + n\,CO_2 \xrightarrow{\text{sunlight}} (CH_2O)_n + n\,O_2$
4. cytosol, mitochondria
5. S
6. acyl CoA
7. liver, acetyl CoA
8. unsaturated
9. chloroplast,
   light, ATP, NADPH, H^+, O_2
   ATP, NADPH, H^+, carbohydrate
10. cytoplasm
11. nonessential
12. isopentenyl pyrophosphate (C_5), geranyl pyrophosphate (C_{10}), farnesyl pyrophosphate (C_{15})

"I hate quotations. Tell me what you know."
- Ralph Waldo Emerson

29

Nutrition

CHAPTER OBJECTIVES

After you have studied the chapter and worked the assigned exercises in the text and the study guide, you should be able to do the following.

1. Define the term, nutrient, and classify the nutrients into the six groups.
2. Indicate what RDA and DRI stands for.
3. Explain the difference between basal caloric requirement and the normal caloric intake of active individuals.
4. Explain why a lengthy diet is usually necessary to lower weight by more than a few pounds.
5. Indicate the primary roles in the body for carbohydrates, fats and proteins.
6. Distinguish between digestible and non-digestible carbohydrates.
7. Distinguish between a complete and an incomplete protein from a nutritional standpoint.
8. Outline indicate what occurs during the process of digestion.
9. Outline the digestion process for carbohydrates and point out the role of the most important enzymes involved.
10. Outline the digestion process for lipids and point out the role of the most important enzymes involved.
11. Outline the digestion process for proteins and point out the role of the most important enzymes involved.
12. Name the water-soluble vitamins and indicate one function for each.
13. Name the fat-soluble vitamins and indicate one function for each

14. Define the important terms and comparisons in this chapter and give specific examples where appropriate.

IMPORTANT TERMS AND COMPARISONS

Nutrients: Carbohydrates, Lipids, Proteins,
    Vitamins, Minerals and Water
Nutritional Calorie (Cal) = 1 kcal
Discriminatory Curtailment Diets
Basal Calorie Requirement
α–Amylase, β–Amylase and
    the Debranching Enzyme
Lipids and Lipases
Essential Fatty Acids: Linolenic and Linoleic Acids
Complete Protein
Protein Complementation
Kwashiorkor
Fat-Soluble Vitamins
    Vitamin A
    Vitamin D
    Vitamin E
    Vitamin K
Vitamin and Corresponding Coenzyme
Minerals
Artificial Sweeteners
Performance Enhancing Foods: Creatine
Anabolic Steroids
Olestra
Hypoglycemia and Hyperglycemia

Digestion
Dietary Reference Intakes (DRI) and
    Recommended Daily Allowance (RDA)
Fiber and Cellulose
Marasmus and Obesity
Body Mass Index (BMI)
Starch and Glycogen
Bile Salts
Essential Amino Acids
Digestive Proteolytic Enzymes
    Chymotrypsin, Pepsin
    Trypsin, Carboxypeptidase
Water-Soluble Vitamins
    Vitamin C (Ascorbic Acid)
    Vitamin B_1 (Thiamine)
    Vitamin B_2 (Riboflavin)
    Vitamin B_6 (Pyridoxal)
    Nicotinic Acid (Niacin)
    Folic Acid
    Vitamin B_{12}
    Pantothenic Acid
    Biotin
Parenteral Nutrition

SELF-TEST QUESTIONS

MULTIPLE CHOICE. In the following exercises, select the correct answer from the choices listed. More than one correct answer may be possible in these questions.

1. Which of the following are classified as nutrients?

    a. carbohydrates
    b. nucleic acids
    c. esters
    d. vitamins
    e. water

2. A nutrient calorie (Cal.) is equivalent to

    a. 100 calories
    b. 1 calorie
    c. 1000 calories
    d. no fixed number because it will depend on the route of digestion

3. The most prevalent form of dietary fats is (are)

a. fatty acids
b. cholesterol
c. triglycerides
d. sphingolipids

4. The essential fatty acids are

a. linolenic and linoleic acids
b. acetic and linoleic acids
c. palmitic and linolenic acids
d. stearic and linoleic acids

5. Water makes up what percent of body weight

a. 15%
b. 75%
c. 35%
d. 60%

6. A number of enzymes are necessary for the digestion of carbohydrates. The role of β-amylase is to

a. hydrolyze the α (1-4) glycosidic linkages in an orderly fashion, cutting disaccharide maltose units one by one from the nonreducing end of the chain
b. randomly cut the α (1-4) glycosidic linkages
c. hydrolyze the α (1-6) glycosidic linkage
d. hydrolyze all glycosidic linkages, irrespect of the type of linkage

7. A carbohydrate that cannot be hydrolyzed in our digestive system is

a. glycogen
b. cellulose
c. amylose
d. amylopectin

8. Some of the final digestion products of lipids as a result of lipase action are

a. glycerol
b. choline
c. phospholipids
d. sphingolipids

9. Digestion of proteins results from the hydrolysis of the

a. phosphodiester linkage
b. peptide linkage
c. glycosidic backbone
d. hemiacetal linkage

10. Enzymes involved in the digestion of proteins include

a. pepsin
b. lipase
c. carboxypeptidase
d. α-amylase

11. The water-soluble vitamins include

a. vitamin B_{12}
b. biotin
c. thiamine
d. niacin

12. The only fat-soluble vitamin in the following list is

a. vitamin K
b. niacin
c. ascorbic acid
d. folic acid

COMPLETION. Write the word, phrase or number in the blank space in answering the following questions.

1. In order to lose 10 pounds of fat, a person must use up __________ Cals.

2. We must consume certain amounts of vitamins, minerals and nutrients. Typically, we must take in ___________ (a quantity) of nutrients, while we need only __________(quantity) of vitamins and minerals.

3. Most of the minerals are metal ions. Five of the metal ions are ______________, ______________, ________________, _________________, and _______________.

4. Two minerals that are not metals are ________________ and _________________.

5. Two fat-soluble vitamins are _________________ and ________________.

6. Four water soluble vitamins are _______________, ______________, ____________ and ______________.

7. Linolenic and linoleic are essential fatty acids in higher animals. They both contain ______________ carbons in the make-up of the fatty acid.

8. Three water soluble vitamins that are converted to coenzymes in biological reactions are _______________, _________________ and ________________.

9. _________________ is a complete protein, while examples of proteins that are not complete are _______________ and ________________.

10. _______________ are the most concentrated form of energy that we consume as food.

11. Olestra is effectively a fat since is has ____________ functional groups. It is derived from the sugar, ___________, in which the ____ groups have been esterified by reaction with long chain fatty acid groups.

12. Although __________ is not a nutrient, it is an important indigestible ingredient in our diets.

MATCHING. Match the term in the right-hand column with the appropriate answer in the left-column.

1.	Vitamin A	a.	Fat-soluble antioxidant
2.	Thiamine	b.	Found in leafy vegetables
3.	Ascorbic Acid	c.	Contains the metal, cobalt
4.	Vitamin K	d.	Another name for vitamin B_1
5.	Vitamin E	e.	Important in bone formation
6.	Vitamin D	f.	Prevalent in citrus fruit
7.	Folic Acid	g.	Important for blood clotting
8.	Vitamin B_{12}	h.	Important for proper vision

ANSWERS TO SELF-TEST QUESTIONS

Multiple Choice

1. a, d, e
2. c
3. c
4. a
5. d
6. a
7. b
8. a, b, d
9. b
10. a, c
11. a, b, c, d
12. a

Completion

1. 35,000 Cal.
2. (many) grams; micro- or milligrams
3. Any five of Na, Ca, Mg, Fe, Zn, Cu, Mn, Cr, Mo or Cu
4. Any two of Cl, P, Se, I or F.
5. Any two of vitamins A, D, E or K.
6. Any four of vitamins B_1, B_2, B_6, B_{12}, niacin, folic acid, pantothenic acid, biotin, or vitamin C.
7. 18
8. Any 3 of vitamins B_1, B_2, B_6, B_{12}, niacin or folic acid.
9. Casein, any two of the following; gelatin or corn, rice or wheat protein.
10. Fats
11. ester, sucrose, -OH
12. fiber

Matching

1. h
2. d
3. f
4. g
5. a
6. e
7. b
8. c

"I hate quotations. Tell me what you know."
- Ralph Waldo Emerson

30

Immunochemistry

CHAPTER OBJECTIVES

After you have studied the chapter and worked the assigned exercises in the text and the study guide, you should be able to do the following.

1. Indicate the characteristics that distinguish innate immunity and acquired immunity.
2. Outline the role of B cells in the development of an immune response.
3. Outline the role of the T cells in the development of an immune response.
4. Indicate the characteristics of killer and memory T cells.
5. Describe the interaction between an antigen and an antibody, emphasizing the molecular interactions involved on each.
6. Describe the function of Class I and Class II major histocompatibility complexes (MHC) in the immune response.
7. Distinguish between the different classes of immunoglobulins in terms of size, function and location.
8. Characterize the variable and constant regions of an immunoglobulins in terms of function and the region within the protein.
9. Outline three different ways that create mutations in the V(D)J genes that help to explain antibody diversity.
10. Characterize the different types of cytokines
11. Define the important terms and comparisons in this chapter and give specific examples where appropriate.

IMPORTANT TERMS AND COMPARISONS

Innate Immunity - External and Internal
Immune System: Specificity and Memory
Antibodies and Immunoglobulins
Lymphocytes - B Cells and T Cells
Antigens and Antibodies
Cytokines - Signaling Regulatory Proteins
Macrophages and Natural Killer Cells
NO and NO Synthase
Plasma Cells
Antibodies and Immunogens
Major Histocompatibility Complex (MHC)
Classes of Immunoglobulins
Large and Small Chain (Subunits)
Antigen-Antibody Complex
Variable Region in H Chain and
V(D)J Recombination
Monoclonal Antibodies
Glycoprotein 120 (gp120)
Interleukins and Chemokines
T Cell Receptor Complex
Activating and Inhibitory Receptors

Acquired or Adaptive Immunity
Immunoglobulin Superfamily
Antibodies, T Cell Receptors and
Major Histocompatibility Complexes
(MHCs)
Lymph and Lymphoid Organs
Specific and Non-Specific Interactions
T Cell Differentiation:
Killer and Memory T Cells
Epitope on an Antigen
MHCs - Class I and Class II
IgG, IgM, IgA, IgD and IgE
Glycoproteins
Monoclonal Antibodies
Cluster Determinant (CD) Proteins
T-Cell Receptor (TcR)
Immunization
Adhesion Molecules: CD4 and CD8
HIV Virus and AIDS
Autoimmune Diseases
Glycomics

SELF-TEST QUESTIONS

MULTIPLE CHOICE. In the following exercises, select the correct answer from the choices listed. More than one correct answer may be possible in these questions.

1. The initial binding of a foreign molecule is carried out by

   a. macrophages
   b. CD4
   c. MHC
   d. B cells

2. An epitope is found on

   a. an immunoglobulin protein
   b. a T-cell surface protein
   c. an antigen
   d. a cytokine

3. The role of the MHC is to

   a. bind CD4
   b. stimulate cytokine formation
   c. bind to antigen in plasma
   d. bringing the antigen's epitope to the cell surface

4. The smallest (by MW) antibody in the blood is

   a. IgG
   b. IgD
   c. IgE
   d. IgM

5. The number of polypeptide chains in an immunoglobulin is

a. one
b. four
c. two
d. six

6. The region of the polypeptide chains in antibodies that interacts directly with the antigen is

a. N-terminal end
b. C-terminal end
c. central region only
d. distributed over many regions

7. B cells are stimulated to differentiate into plasma cells by

a. exposure to cytokines
b. antigen binding to the surface-bound antibody
c. on interaction with macrophages
d. after MHC display on surface

8. Each B cell is capable of synthesizing

a. many different antibodies
b. an unknown number that depends on the situation
c. only one immunoglobulin
d. more than one, but clearly a limited number

9. T cells bind to antigen by their surface

a. TcR complex
b. TcR and cytokine together
c. TcR
d. CD4 receptor

10. Monoclonal antibodies are derived from

a. a single B cell
b. multiple B cells
c. a single T cell
d. a specific type of macrophage

11. Cytokines are proteins that are modified by

a. sugar groups (glycoproteins)
b. acetyl groups
c. phosphate groups
d. methyl groups

12. The proteins that have an epitope that an antibody binds to are the

a. macrophages
b. MHCs, class II
c. antigens
d. CD4 proteins

COMPLETION. Write the word, phrase or number in the blank space in answering the following questions.

1. Myasthenia gravis is an __________ disease in which the patient develops antibodies against ___________ receptors.

2. Heceptin is a _____________ antibody that is used in the treatment of breast cancer and binds to the antigen called _________.

3. A immunization shot produces only a ____ ____ ____ and is not effective. If the virus infects the cells or alternatively, a second immunization shot, the so-called "booster shot", is given, this increases the response time and its effective because the ________ cells divide and become both _________ and _________ cells.

4. A key difference between innate and acquired immunity is that the former exhibits no ______ or __________, while these characteristics are the hallmark of acquired immunity.

5. Of the two major types of T cells, only the ________ T cells release the protein, ________ , that effectively plugs a hole in the target cell.

6. The _____ cells are derived from ____ cells, which produce ______ proteins.

ANSWERS TO SELF-TEST QUESTIONS

Multiple Choice

1. a
2. c
3. d
4. a
5. b
6. a
7. b
8. c
9. a
10. a
11. a
12. c

Completion

1. autoimmune, acetylcholine
2. monoclonal, c-erb-B2
3. short term response, memory, memory, plasma
4. specificity, memory
5. killer, perforin
6. plasma , B, antibodies

"The thing to remember when traveling is that the trail is the thing, not the end of the trail. Travel too fast and you miss all you are traveling for."
- Louis L'Amour

31

Body Fluids

CHAPTER OBJECTIVES

After you have studied the chapter and worked the assigned exercises in the text and the study guide, you should be able to do the following.

1. Tabulate the functions of blood and indicate which molecules play a major role in these functions.
2. Characterize each of the three main blood cell types, including where they are formed and their major functions.
3. Compare the components of blood, blood plasma, and serum.
4. Discuss the interrelationship between prothrombin, thrombin, thromboplastin, and thrombosis.
5. Outline the characteristics of the hemoglobin molecule, its O_2 binding characteristics and how the CO_2 and H^+ influence O_2 binding.
6. Explain the role of nephrons in retaining proteins and vital nutrients, while eliminating waste products from the blood.
7. Discuss the role of the hormones vasopressin and aldosterone in the control of water and salt balance in the blood and kidneys.
8. Outline how blood pressure is regulated normally with diet and hormonal controls.
9. Define the important terms and comparisons in this chapter and give specific examples where appropriate.

IMPORTANT TERMS AND COMPARISONS

Extracellular, Interstitial and Intercellular Fluid
Erythrocytes and Red Blood Cells
Platelets and Thrombocytes
Edema
Blood-Brain Barrier
Hemoglobin and Red Blood Cells
Bohr Effect
Dalton's Law of Partial Pressures: O2 and CO2
Oxygen Dissociation Curve
Kidney Function: Blood Cleansing
Proximal Tubule, Henle's Loop and Distal Tubule
Fibrinogen and Fibrin
Baroreceptor
Renin, Angiotensinogen and Angiotensin
Myocardial Infarction
Prostate-Specific Antigen (PSA)
Water and Salt Balance: Vasopressin and Aldosterone
LeChatelier's Principle
Leukocytes and White Blood Cells
Whole Blood, Blood Plasma and Serum
Acidosis and Alkalosis
Homeostasis
Allosteric Effect and Cooperative Binding
Arterial and Venous Blood
Carbaminohemoglobin
Blood Clotting and Anticoagulant Drugs
Nephrons and Bowman's Capsule
Glomeruli
Urine, Diuresis and Antidiuretic
Prothrombin and Thrombin
Hypertension
Diuretics
Carbonic Anhydrase
Vasoconstrictor

SELF-TEST QUESTIONS

MULTIPLE CHOICE. In the following exercises, select the correct answer from the choices listed. In some cases, two or more choices will be correct.

1. The plasma contains

   a. albumin
   b. erythrocytes
   c. hemoglobin
   d. globulin

2. The movement of molecules from the blood to the cerebrospinal fluid is limited and regulated by the

   a. blood pressure
   b. blood-brain barrier
   c. renin
   d. O2 pressure in the arteries

3. The Bohr effect is associated with the

   a. blood filtration in the kidneys
   b. action of hormones in the blood
   c. salt balance in the cell
   d. pH effect on O_2 binding to hemoglobin

4. Blood clotting can be prevented during transfusions by the addition of sodium citrate. This agent

   a. complexes with Ca^{+2}
   b. converts thrombin back to prothrombin
   c. produces edema
   d. degrades collagen

5. Extracellular fluids are those

   a. within the cell
   b. not inside the cell
   c. found in cytosol
   d. with little or no Cl^-

6. A molecule of CO_2 binds to hemoglobin to form carbaminohemoglobin. The CO_2 forms a covalent bond at the

   a. Fe(II) in the heme group
   b. amino-terminal end of the polypeptide chain
   c. carboxyl-terminal end of the polypeptide chain
   d. H^+ binding site

7. Carbonic anhydrase catalyzes which of the following reactions.

   a. $Hb + O_2 \rightleftharpoons HbO_2$ Hb = hemoglobin

   b. $CO_2 + H_2O = H_2CO_3$

   c. $HbO_2 + H^+ + CO_2 = Hb(CO_2)H^+ + O_2$

   d. $H_2PO_4^- = H^+ + HPO_4^{-2}$

8. Vasopressin is

   a. a diuretic hormone
   b. an antidiuretic hormone
   c. a small peptide
   d. manufactured in the hypophysis

9. Hemoglobin more readily binds to the second, third and fourth O_2 molecules than it does the first because of

   a. the Bohr effect
   b. an allosteric effect
   c. a pH effect
   d. Henle's loop

10. Some of the functions of the blood include

   a. maintaining the pH of the body
   b. fighting infection
   c. carrying nutrients
   d. carrying O_2 from the lungs to the tissues

11. The cells that are made in the bone marrow include

   a. erythrocytes
   b. leukocytes
   c. platelets
   d. nephrons

12. An effective anticoagulant in the blood

   a. thins the plasma
   b. complexes with Ca^{+2} ions
   c. decreases the number of erythrocytes
   d. reduces the chloride concentration

13. The hemocrit level in a blood sample is the

   a. percent of erythrocytes in the blood
   b. oxygen (gas) pressure
   c. percent of leukocytes in the blood
   d. percent of oxygenated hemoglobin in the blood

COMPLETION. Write the word, phrase or number in the blank space in answering the question.

1. Severe dehydration stimulates the production of the hormone ______________ to enable the body to retain more ______________.

2. Hormonal control of blood pressure involves the secretion of the enzyme ____________ from the kidney. This enzyme converts the inactive protein, ______________, into the potent vasoconstrictor, ______________.

3. The serum contains all the elements of plasma except for ______________.

4. Persons with hypertension have ______________ blood pressure. This can lead to a number of medical problems, including ______________ or ______________.

5. The ______________ molecule is bound to the __________(a specific atom) in the fully oxygenated form of hemoglobin.

6. ______________ is the major component of the blood which is a liquid and contains no ______________.

7. ______________ is an ______________ that initiates blood clotting. It is ordinarily found in the blood in the inactive form called ______________.

8. Most of the CO_2, a waste product in cellular metabolism, is transported to the lungs as ______________.

9. The ______________ organ acts a superfiltration unit, in which the waste products are eliminated in the ______________.

10. ______________ is the name of the condition which occurs when the pH of the blood increases above normal.

11. A nephron contains ______________ capsule which is connected to a tiny tube called the ______________. The U-shaped twist in the unit is called ______________ ______________.

12. The _____ gas that we breathe is transported to the actively metabolizing cells in our bodies by the molecule, __________, which is in the ________ cells in the blood. Once in the cell, this molecule is ______ to water in the mitochondria and generates the energy molecule, ______. This latter process is called _______ ________.

ANSWERS TO SELF-TEST QUESTIONS

Multiple Choice

1. a, d
2. b
3. d
4. a
5. b
6. b
7. b
8. b, c, d
9. b
10. a, b, c, d
11. a, b, c
12. b
13. a

Completion

1. aldosterone, water
2. rennin, angiotensinogen, angiotensin
3. fibrinogen
4. high, heart attack, stroke, kidney failure (any two of these)
5. O_2, Fe in the heme group
6. plasma, cells
7. thrombin, enzyme, prothrombin
8. H_2CO_3
9. kidney, urine
10. alkalosis
11. Bowman's, tubule, Henle's loop
12. O_2, hemoglobin, erythrocyte, reduced ATP, oxidative phosphorylation